T0140831

Structure Performance Relationships of the Novel MALDI-MS Matrices

Zur Erlangung des akademischen Grades eines

DOKTORS DER NATURWISSENSCHAFTEN

(Dr. rer. nat.)

von der KIT-Fakultät für Chemie und Biowissenschaften

des Karlsruher Instituts für Technologie (KIT)

genehmigte

DISSERTATION

von

MSc. TAMBE SUPARNA BHASKAR

Referent: Prof. Dr. Stefan Bräse

Korreferent: Prof. Dr. Burkhard Luy

Tag der mündlichen Prüfung: 16. Dezember 2016

Band 72
Beiträge zur organischen Synthese
Hrsg.: Stefan Bräse

Prof. Dr. Stefan Bräse
Institut für Organische Chemie
Karlsruher Institut für Technologie (KIT)
Fritz-Haber-Weg 6
D-76131 Karlsruhe

Bibliographic information published by the Deutsche Nationalbibliothek

The Deutsche Nationalbibliothek lists this publication in the Deutsche Nationalbibliografie; detailed bibliographic data are available in the Internet at http://dnb.d-nb.de .

ISBN 978-3-8325-4621-2
ISSN 1862-5681

Logos Verlag Berlin GmbH
Comeniushof, Gubener Str. 47,
10243 Berlin
Tel.: +49 030 42 85 10 90
Fax: +49 030 42 85 10 92
INTERNET: http://www.logos-verlag.de

List of Abbreviations

3-AP	3-Acetylpyridine
4-Me-p-PhCCAA	4-Methyl-p-phenyl-α-cyanocinnamic acid amide
9-AA	9-Aminoacridine
ACN	Acetonitrile
AnCCA	(*E*)-3-(Anthracen-9-yl)-2-cyanoprop-2-enoic Acid
BSA	Bovine Serum Albumin
BTLE	Brain Total Lipid Extract
ClCCA	4-Chloro-α-cyanocinnamic acid
DHA	Dihydroxyacetophenone
DHB	2,5-Dihydroxybenzoic acid
Di-FCCA	2,4-Difluoro-α-cyanocinnamic acid
DPDA	2-Cyano-5,5-diphenyl-dienoic acid
EVD	Eigen Value Decomposition
EWG	Electron Withdrawing Group
FAB	Fast Atom Bombardment
HCCA	4-Hydroxy-α-cyanocinnamic acid
HFBA	Heptafluorobutyric acid
HFIP	1,1,1,3,3,3-Hexafluoro-2-propanol
IPA	iso-propylalcohol
IR	Infrared
MALDI	Matrix Assisted Laser Desorption Ionisation
MBT	2-Mercaptobenzothiazole
m-PhCCAA	m-Phenyl-α-cyanocinnamic acid amide
MS	Mass Spectrometry
NIPALS	Non-Iterative Partial Least Squares
NMP	*N*-Methylpyrrolidine

NMR	Nuclear Magnetic Resonance
NpCCA	(*E*)-2-Cyano-3-(naphthalen-2-yl)acrylic Acid
PA	Phosphatidic acid
PC	Phosphatidylcholine
PCA	Principal Component Analysis
PDA	2-Cyano-5-phenyl-dienoic acid
PE	Phosphatidylethanolamine
PG	Phosphatidylglycerol
PI	Phosphatidylinositol
PNA	p-Nitroaniline
p-PhCCAA	p-Phenyl-α-cyanocinnamic acid amide
PS	Phosphatidylserine
SA	Sinapinic acid (3,5-dimethoxy-4-hydroxycinnamic acid
SM	Sphingomyelin
ST	Sulfatide
SVD	Singular Value Decomposition
TFA	Trifluoroacetic acid
THAP	2,4,6-trihydroxyacetophenone
UV	Ultra-Violet

Abstract

MALDI (Matrix Assisted Laser Desorption Ionisation) mass spectrometry is a rapidly developing soft ionisation method that finds its applications in the detection and analysis of biomolecules like proteins, peptides, lipids, carbohydrates, nucleotides etc. and is useful for the identification of bacteria and imaging as well. Although MALDI-MS has diverse applications since its origin, the knowledge of matrix optimisation technique is meager. The issue of what makes a chemical compound good or bad matrix still remains more or less unexplored. It is imperative to ascertain which physical and chemical properties of a chemical compound are vital for its role as a matrix. As a result developing new matrices for MALDI-MS is still a matter of trial and error.

For this research, a chemical compound library of approximately 200 cinnamic acid derivatives was synthesized. These potential matrices were used to detect lipids from brain total lipid extract and peptides from BSA digest in positive and negative ion modes of mass spectrometry. Eventually the results were compared with commonly used MALDI matrices like 4-hydroxy-α-cyanocinnamic acid (HCCA), 2,5-dihydroxybenzoic acid (DHB), 4-chloro-α-cyanocinnamic acid (ClCCA) and/or 9-amino acridine (9-AA).

During this doctoral study, many novel matrices were discovered that exhibit superior qualities, in comparison to the established matrices, for different analyte classes which are presented in the results and discussion section of the thesis. It is worth mentioning that a novel matrix p-phenylcinnamic acid amide (p-PhCCAA) was developed for MALDI-imaging of lipids using 355 nm laser; which displays superior sensitivity and reproducibility to the standard matrix 9-AA for negative ion mode. A small number of background peaks and good matrix suppression effect makes this new matrix widely applicable tool for lipid analysis using 355 nm laser.

A key aspect for the ongoing advancements of MALDI-MS is a profound understanding of the intricacies in working principles of MALDI matrices. In this thesis, organic synthesis was used as a tool for complementing analytical chemistry to gain insight into the relationship between matrix properties, structure and their MALDI-MS performance in a certain analyte class. Cater to this demand, a compound library of 59 structurally related phenylcinnamic acid derivatives was designed and synthesized. Potential MALDI matrices were evaluated with sulfatides, a class of anionic lipids; which are abundant in complex brain lipid mixtures using 337 nm and 355 nm lasers. The generation and analysis of a large data set recorded on two different MALDI-MS instruments using two different lasers (337 and 355 nm) allowed the identification of important structural features for MALDI-Matrix performance other than molar extinction coefficients. The geometry of compounds and the substitution position on the aromatic ring were identified as vital matrix properties.

Using this multidisciplinary approach; it was shown for the first time that for the rational development or selection of matrix, chemical library synthesis, and strategies used in medicinal chemistry and multivariate statistics are viable methods to identify important novel principles. Besides providing a class of novel MALDI-MS matrices the results will facilitate the matrix discovery process for analytical chemists. This is a significant step towards a rule-based design of MALDI-MS matrix compounds.

Kurzzusammenfassung

Die MALDI (Matrix Assisted Laser Desorption Ionisation) Massenspektrometrie ist eine sich schnell entwickelnde spektrometrische Technik, die eine schonende Ionisation ermöglicht. Anwendung findet sie in der Detektion und Analyse von Biomolekülen (Proteine, Peptide, Lipide, Kohlenhydrate, Nucleotide etc.), sowie zur Identifizierung von Bakterien und im Bereich der Bildgebenden verfahren. Trotz der schnellen Entwicklung der MALDI-MS ist noch wenig bekannt über die Matrix-optimierung. Es ist nicht klar, welche chemischen oder physikalischen Eigenschaften einer Verbindung die Qualität einer Matrix beeinflussen. Bedingt durch diesen Mangel an Informationen werden MALDI-MS Matrices immer noch nach dem Prinzip "trial and error" entwickelt.

In diesen Arbeit wurde eine Substanzbibliothek bestehend aus ungefähr 200 Zimtsäurederivaten synthetisiert. Diese potentiellen Matrices wurden verwendet um im positiven und negativen Mass-spektrometrischen Ionenmodus sowohl Lipide als auch BSA Verdaue zu untersuchen. Die messergebnisst wurden mit anderen häufig eingesetzten Matrices wie 4-hydroxy-α-cyanocinnamic acid (HCCA), 2,5- dihydroxybenzoic acid (DHB), 4-chloro-α-cyanocinnamic acid (ClCCA) und 9-aminoacridine (9-AA) verglichen.

Im Laufe dieser Doktorarbeit wurden viele neue Matrices für die Anwendung mit unterschiedlichen Klassen von Analyten entwickelt, die Ergebnisse werden im Ergebnis-/ Diskussionsteil erläutert. Dabei wurde eine neue Matrix p-Phenylzimtsäureamid (p-PhCCAA) identifiziert, die für das Lipid Imaging, unter Verwendung eines 355nm Lasers im negativen Ionenmodus, eine erhöhte Empfindlichkeit und Reprozierbarkeit im Vergleich zur Standardmatrix 9-AA zeigt. Ein geringer Background und eine gute Matrix Suppression erlauben einen breitgefächerten Anwendungsbereich für die Lipid Analytik unter Verwendung eines 355 nm Lasers.

Eine Schlüsselrolle für die weitere Entwicklung der MALDI-MS ist ein bessere Verständnis der Wirkprinzipien der MALDI Matrices. Um die Beziehungen zwischen der Struktur, den Matrixeigenschaften und der MALDI-MS Leistungsfähigkeit für eine bestimmte Klasse von Analyten zu vergleichen wurde die organische Synthese als ein Werkzeug der Analytischen Chemie genutzt. Es Aspekt wurden 59 strukturell ähnliche Phenylzimtsäurederivate hergestellt und evaluiert um die Leitstrukturen der potentiellen Matrices zu identifizieren. Die potentiellen Matrices wurden unter Verwendung von Sulfatiden, einer Klasse anionischer Lipide, die als komplexes Gemisch im Cerebralgewebe vorkommen, mit einem 335 nm und 337 nm Laser untersucht, um ihre Leistungsfähigkeit zu bewerten. Die MALDI-MS Messungen im Vergleich zu einem Lipidstandard aus Cerebralgewebe erlaubten genaue Untersuchungen der zu Grunde liegenden Zusammenhänge zwischen Struktur und Leistungsfähigkeit der potentiellen Matrices. Ein umfangreicher Datensatz, welcher unter Verwendung zweier MALDI Massenspektrometer mit zwei verschiedenen Lasern (337nm und 355nm) generiert wurde, erlaubte die Identifizierung wichtiger Substanzeigenschaften, der molaren Extinktionskoeffizienten ist nicht das einzige Kriterium. Sowohl die Molekülgeometrie als auch das Substitutionsmuster des aromatischen Ringes sind wichtige Matrix-Eigenschaften.

In diesem multidisziplinären Ansatz wurde das erste Mal gezeigt, dass die Synthese von Substanzbibliotheken, eine Strategie, die Verwendung in der Medizinischen Chemie findet, in Kombination mit multivariater Statistik zur Identifizierung wichtiger, neuer und signifikanter Prinzipien führt, was eine rationale Entwicklung und Auswahl neuer Matrices ermöglicht. Außerdem vereinfachen diese Ergebnisse den Entwicklungsprozess für den Analytiker und somit einen Schritt in Richtung regelabhängiger Struktur Wirkungsbeziehungen.

Contents

1

Introduction

Matrix assisted laser desorption ionisation (MALDI) is a soft ionisation mass spectrometric technique and was invented by Tanaka [1], Karas, Hillenkamp [2] and his co-workers [3]. MALDI-MS has developed rapidly in its applications that range from detection and identification of biomolecules like peptides, proteins [4], oligonucleotides [5], polysaccharides [6], glycoproteins [7] and lipids [8] to disease diagnosis [9]. MALDI has established a great impact on biomedical [10], chemical [11] and pharmaceutical [12] industries.

1.1 Historical Background of MALDI-MS

Lasers are used to generate ions for mass spectrometry since 1960s [13; 14; 15]. Although low mass molecules and laser absorbing organic molecules are easily accessible by the Laser Desorption Ionisation (LDI), spectra for biomolecules were difficult to obtain and needed some efforts [16]. The first systematic attempt to generate ions from laser desorption ionisation with the organic molecules started from early 1970s [16], which after several observations led to two general principles: 1) efficient and controllable energy transfer is possible using far UV or far IR and 2) the energy must be transferred in very short time to avoid thermal decomposition of thermally labile organic molecules. Earlier, there was an upper limit on the size of molecules that can be desorbed without fragmentation as intact ions, 1000 Da for biopolymers and 9000 Da for synthetic polymers [17]. Until the end of 1980s, fast atom bombardment (FAB) and californium plasma desorption (^{252}Cf-PD) were more effective to generating high quality spectra for the biological samples, as compared to the LDI [18]. The introduction of light absorbing compounds added during the sample preparations for laser desorption mass spectrometry finally effected major change. Two approaches were developed 1) the admixture of ultrafine cobalt powder (particle size of 30 nm) to the analyte solutions in glycerol [1; 19]. 2) The cocrystallisation of an organic compound (matrix) along with the analyte [2; 20]. The use of matrix leading to low fragmentation of higher mass molecules was major breakthrough that led to the concept of MALDI [20]. The versatility of matrix and and sensitivity of MALDI technique [20; 21] made it by far superior to the ultrafine cobalt powder admixture in liquid matrix [22].

1.2 Principle of MALDI-MS

The Figure 1.1 shows schematically the functioning of the MALDI process. The matrix and analyte are mixed using suitable solvent on a target plate. The solvent is dried, leaving behind analyte molecules homogeneously dispersed with matrix molecules. When a laser beam of appropriate wavelength is shot, the laser energy is absorbed by the matrix molecules. Some of this laser energy is transferred to the analyte molecules. This causes desorption of matrix and intact analyte molecules from the target surface into vapour phase. After ionisation, they are forced to pass through the time of flight (TOF) tube by high voltage and in high vacuum. In the TOF tube the distribution of molecules emanating from a sample is imparted to identical translational kinetic energies after being subjected to the same electrical potential difference.

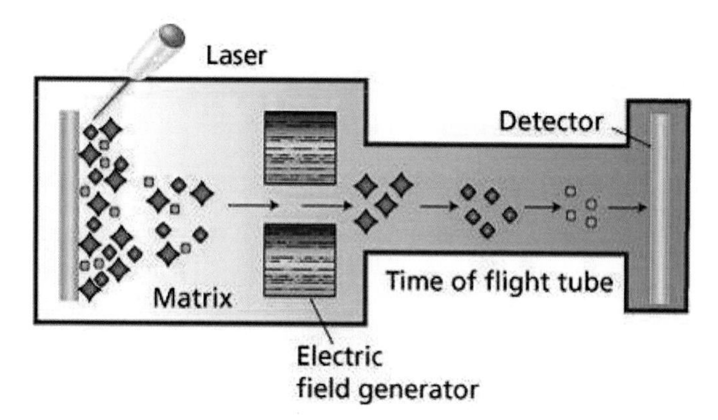

Figure 1.1: Schematic representation of MALDI-MS working principle adopted from http://www.sigmaaldrich.com

The ions while travelling identical distance through the evacuated drift tube, smaller ions reach the detector in shorter time compared to the heavier ions. The digitized data generated from successive laser shots add up to give a MALDI spectrum.

1.2.1 Ion Sources for LDI and MALDI

Absorbtion of laser light energy is used in both the LDI and MALDI techniques. Wavelengths ranging from UV to IR have been employed using nitrogen lasers (337 nm), excimer lasers (193, 248, 308 and 351 nm),Q-switched, frequency-tripled and quadrupled Nd:YAG lasers (355 and 266 nm, respectively) [23], Er:YAG lasers (2.94 μm) [23; 24] and TEA-CO_2 lasers (10.6 μm)[25].

The pulse of laser light is focused onto a small spot (0.05−0.2 mm in diameter) [26]. As laser irradiance is a critical parameter, a variable beam attenuator is employed to adjust the irradiance e.g., by means of a rotating UV filter of variable transmittance from close to 100% down to about 1%. Then, the laser attenuation is individually optimized for each measurement. These ion sources are generally operated at ambient temperature.

UV lasers emit pulses of 3−10 ns duration, while those of IR lasers are in the range of 6−200 ns. Sudden ablation of the sample layer is caused by the short pulses. In addition, an extremely short time interval of ion generation avoids thermal degradation of the analyte. Longer irradiation would simply cause heating of the bulk material. In case of IR-MALDI, a slight decrease in threshold fluence has been observed for shorter laser pulses [27].

1.2.2 Ionisation of Molecules

Zenobi stated a very important observation [28] as follows regarding the ionisation mechanism: "Enhanced knowledge of the ion formation process could provide rational guidelines for matrix selection for a given analytical problem." The MALDI ionisation process is the key step that is still not so well understood [28; 29; 30]. The ionisation mechanism is a subject to continue research [28; 31; 32] The major concerns are the relationship between ion yield and laser fluence [26; 33], the temporal evolution of the desorption process and its implications upon ion formation [34], the initial velocity of the desorbing ions [27; 35; 36], and which ions are the major source of the ionic species detected by MALDI-preformed ones or those which are generated in the gas phase? [37; 38].

The general explanation of ionisation mechanism was provided [39] as follows. The ionisation is a two step process in which initial primary ionisation is followed by the secondary ionisation in the dense plume. The variations in the experimentally accesible parameters such as laser fluence and irradiation wavelength together with some matrix and analyte properties led to really a large number of proposals [28; 29; 32; 39; 40; 41; 42; 43; 44; 45; 46; 47; 48; 49; 50; 51; 52; 53; 54; 55] The primary ionisation process is explained by two different models *viz*. the cluster model [51] and the photoexcitation/pooling model [40; 56].

1.2.3 Ion Yield and Laser Fluence

The fluence is defined as energy per unit area; in MALDI typical fluences are in the range of 10−100 mJ cm^{-2}. The irradiance is fluence divided by the laser pulse duration; in MALDI the irradiances are in the range of 10^6 −10^7 W cm^{-2} [32]. No ion production is observed below the threshold of irradiance at 10^6 Wcm^{-2}. At threshold, a sharp onset of desorption-ionisation takes place and ions abundances rise to 10^5 to 10^9 molecules of the laser irradiance.

1.2.4 Primary Ionisation

There are several mechanisms suggested for the primary ionisation [57; 58; 59; 60; 61; 62; 63; 64] which are based on thermal and droplet models. The most accepted models are discussed in brief in the following sections.

1.2.4.1 The 'Lucky Survivor' Model of Primary Ionisation

This model suggested that the analyte ions are already formed in the solution with the matrix. During ablation, most of the ions recombine with the counterions, while few of the ions escape this recombination called as 'lucky survivors' [36; 51] and are thus detected

1.2.4.2 The Photoionisation/Pooling Model of Primary Ionisation

This model suggested that the ionisation occurs when absorbed energy migrates into the matrix and this energy focuses when pooling event takes place [40; 56]. The Figure 1.2 shows how electrons are excited from S_0 to excited singlet state S_1 or even higher excited states S_n, when the photons are absorbed by matrix molecules during laser excitation. When the two excited states interact with each other, a pooling event takes place. For example, when two neighbouring molecules have S_1 states, after the pooling, one molecule relaxes to S_0 while the other molecule will excite to higher energy state S_n by absorbing energy released by first molecule (shown in Figure 1.2a).

Ionisation takes place when there are two neighbouring molecules with S_1 and S_n states (shown in Figure 1.2b). This pooling will generate S_0 state molecule and an ion [56].

1.2.5 Secondary Ionisation

After formation of ions, they undergo bimolecular collisions with the excess of matrix ions to further undergo protonation, electron emmision or a cation transfer between analyte (A) and matrix (M) ions. The charged ions those can survive these collisions can be detected [39]

$$2M \rightarrow M^{\cdot +} + M^{\cdot -}$$

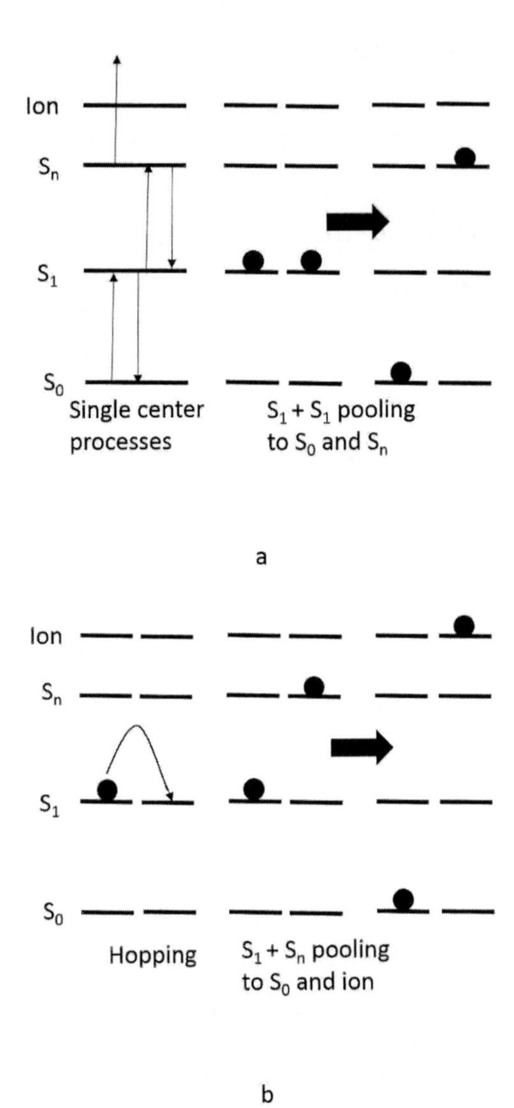

Figure 1.2: MALDI ionisation model proposed for a) unimolecular and b) bimolecular matrix ionisation processes. Adapted from [28]

$$2M \rightarrow MH^+ + (M\text{-}H)^-$$

In presence of analyte a charge transfer reaction takes place

$$M^{\cdot +} + A \leftrightarrows M + A^{\cdot +}$$

1.2.5.1 Matrix Suppression Effect

The signals of matrix ions can be completely suppressed by adjusting matrix-analyte concentration ratios. The mass spectrum will contain only analyte signals. This phenomenon is called as matrix suppression effect (MSE). MSE occurs only when all the matrix ions formed in the primary ionisation process react with the analyte molecules. Thus it is a concentration dependent phenomenon. The concentration of ions, in turn, depends on the laser fluence [39]

1.2.5.2 Analyte Suppression Effect

This occurs when there is a mixture of two or more analytes. For example, consider a mixture of analytes A and B. If A interacts more efficiently with the matrix ions than B and if there is charge transfer between A and B then the signal of analyte B would be weaker in absence of analyte A. This phenomenon is called as analyte suppression effect (ASE) [39].

$$M^{\cdot +} + A \rightarrow M + A^{\cdot +}$$
$$M^{\cdot +} + B \rightarrow M + B^{\cdot +}$$
$$A^{\cdot +} + B \rightarrow A + B^{\cdot +}$$

1.2.5.3 Other Secondary Reactions

One of the most important reactions for analysis of proteins and peptides is proton transfer which is endothermic reaction [39].

In case of synthetic polymers cationisation is the main reaction. The proton transfer is exothermic, in general, and ionisation is inhibited by addition of cation adducts to analyte-matrix mixture, if the polymer has higher affinity to any specific cation compared to that of the matrix [39]

Electron transfer is another secondary reaction for the less polar molecules. In this case, analyte ions are detected as deprotonated species and radical anions [65]

1.2.6 Sample Preparation

Sample-matrix preparation procedures are one of the greatest influencing factors for the quality of MALDI mass spectra [66]. The most common procedure is called 'dried droplet' [66]. In this method, a saturated matrix solution is mixed with the analyte solution such that the matrix to sample ratio of 5000:1 is obtained. Then 0.5-2.0 µL of this mixture is applied on the target and allowed to dry over air.

In the 'thin layer' method [67; 68], homogeneous matrix film is generated on the target in first step and then the analyte is applied and is absorbed by the matrix. It provides good sensitivity, resolving power and mass accuracy [66].

The 'thick layer' method makes use of nitro cellulose as a matrix additive. First nitro cellulose-matrix layer is formed on the target and then the analyte is applied. This method suppresses alkali adduct formation and increases the detection sensitivity [66].

The 'sandwich' method [7] is similar to the thin layer method with some additional steps. First matrix layer is formed followed by applying TFA (trifluoroacetic acid), then analyte and then again a layer of matrix.

1.2.7 The Rate of the Matrix Crystal Growth

The time given for the crystal to grow changes the size of the crystal and it affects the performance. [69]

1.2.7.1 Moderately Fast Crystallisation

The most commonly used procedures for the analyte-matrix sample preparation involves the variations of dried droplet [2] method discussed in section 1.2.6.

1.2.7.2 Slow Crystallisation

The sample-matrix solution is prepared in a micro-centrifuge tube and the solvent is allowed to evaporate very slowly through a hole made on the cap of the tube. The crystal that are formed on the wall of the tube are scraped off and placed on the target plate [70]

1.2.7.3 Rapid Crystallisation

The MALDI matrix-analyte solution is evaporated under vacuum within few seconds. Peptides and proteins analysed by this method normally exhibit extensive alkali cation adduction [69].

1.3 MALDI-MS-Imaging (MSI)

Imaging using mass spectrometry is a new and rapidly growing. MALDI-MSI directly delivers highly parallel and multiplexed data on the specific localisation of molecular ions in tissue samples and is used to measure and map the variations of these ions during development and disease progression or treatment. It has an intrinsic potential to identify several biomarkers in the same experiment by relatively simple extension of the method [71].

The MALDI-MSI is a collection of a mass spectrum at each of a regular series points across a section of tissue. Subsequently, these spectra can be used to plot relative intensity of the individual *m/z* peaks and therefore, it is possible to visualise the distribution of the individual molecular ions [71].

The advantages of MALDI-TOF-MSI are [72]

- A number of biomolecule analyte classes can be measured from different mass ranges, like lipids , drugs, peptides and proteins
- No preliminary knowledge about the tissue composition is necessary

The basic workflow for the MALDI-MSI is shown in Figure 1.3

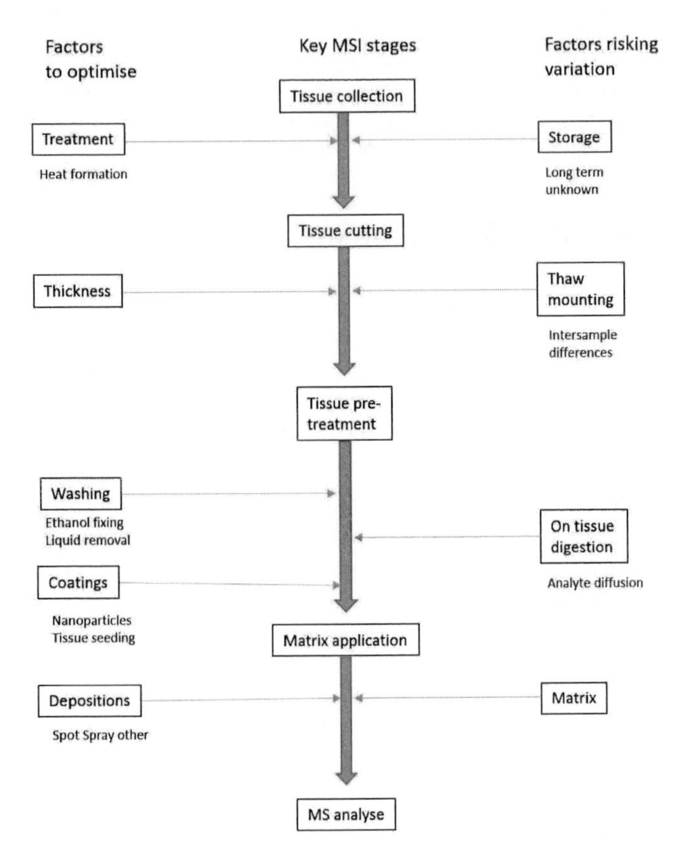

Figure 1.3: Schematic representation of MALDI-MSI workflow showing key steps and the factors needed to be considered to minimise variation and optimise performance (adapted from [71]

For each step in the workflow there are several alternatives, modifications and additions [71] and discussing each step in detail would be out of the scope of this report. For this report the most important step is the correct choice and right method of application of matrix and would be discussed in section 1.3.1 in greater detail.

1.3.1 Choice and Application of the Matrix

The high quality of MALDI-MSI spectra is a direct result of MALDI matrix deposition on the sample. The method of matrix application must ensure the extraction of maximum amount of analyte from the sample but at the same time it must ensure minimum diffusion and homogeneous crystallisation of the matrix. This is possible if correct matrix is applied in correct solvent composition that provide enough humidity for the extraction of analytes but avoids excessive diffusion.

For MALDI-MSI, the key is to generate matrix crystals that are small enough to be below the resolution of the imaging itself and at the same time it should ensure good inclusion of analyte into the crystals while causing less diffusion [71]. Generally for low mass peptides MALDI-MSI is carried out with reflectron mode using TOF mass spectrometer using HCCA (4-hydroxy-α-cyanocinnamic acid) as matrix while for high mass proteins the matrix used typically is sinapinic acid (3,5-dimethoxy-4-hydroxycinnamic acid, SA) [71]. Different modifications of HCCA and SA are shown in table 1.1. The ionic liquid matrix mixtures like HCCA/2-amino-4-methyl-5-nitropyridine and HCCA/N,N-dimethylaniline have been reported for better spectral quality in terms of many factores like resolution, sensitivity, noise, intensity and the number of species detected [71].

Matrix	Biomolecule specificity
2,5-dihydroxybenzoic acid (DHB)	Lipids, Peptides, < 1-kDa proteins
4-hydroxy-α-cyanocinnamic acid (HCCA)	Peptides, < 10kDa proteins
DHB/ aniline	Lipids, peptides, < 10kDa proteins
DHB + 3-acetyl pyridine (DHB/3-AP)	Lipids, peptides, < 10kDa proteins
9-aminoacridine (9-AA)	Lipids
3,5-dimethoxy-4-hydroxycinnamic acid (SA)	> 10kDa proteins
SA/ aniline	> 10 kDa proteins
SA + 1,1,1,3,3,3-hexafluoro-2-propanol (SA/ HFIP)	> 30 kDa proteins

Table 1.1: Matrix and their biomolecular specificities

1.4 Lipids and Their Analysis

The analysis of lipids continues to gain importance by leaps and bounds, such that it has become an -Omics i.e. lipidomics [73]. Lipids define the topology of the cell and its organelles and are the main components in the membranes [74]. They are the most common biomolecules in the brain and they account for almost half of the total dry weight of brain [75]. Lipids perform various roles in the body, for example, storage of energy, signalling across biological membranes and storage of dugs and certain types of organic molecules [76]. The imporatance of the lipids can be pointed out by the fact that the altered levels of phospholipids in tissue are associated with Farber disease, Gaucher disease, Niemann-Pick disease, Alzheimer disease and Down syndrome [76; 77; 78; 79; 80]. Some lipids like sphingomyelin and glycolipids form noncovalent complexes with chlorisondamine, a nicotinic antagonist [81; 82]. The brain tissue lipidomics is important due to the fact that it allows precise anatomic localisation of different lipid species in different areas of the brain [76]. Many therapeutic and abusive drugs localise in certain brain areas and interact with lipids in the membranes noncovalently [76]. These interactions play an exteremly important role in drug delivery and drug effects because these interactions control the rate of release of drugs [76].

1.4.1 Types of Brain Lipids

The brain lipids mainly consists of three categories of lipids *viz.* cholesterol, sphingolipids (sphingomyelin, cerebrosides, sulfatides and gangliosides) and glycerophospholipids (phosphatidylcholine, phosphatidylethanolamine, phosphatidylinositols) [83]. Different types of lipids are shown in Figure 1.4. The figure illustrates the structure of some common classes of glycerophospholipids *viz.* phosphatidylcholine (PC), phosphatidylethanolamine (PE), phosphatidylinositol (PI) and phosphatidylserine (PS) as well as some sphingolipids *viz.* sphingomyelin (SM), sulfatide (ST) and ganglioside. The different head groups of each lipid class affect the ionisation efficiency of the lipids by MALDI-MS [76]. The PCs and SMs are ionised in positive ion mode with ease whereas, PIs, PSs and STs are easily ionised in the negative ion mode of MALDI-MS [76]. PEs are easily ionised in both the positive and negative ion modes of MALDI-MS [76]. The nomenclature for PC, PE, PI and PS species is as follows: the numbers (x:y) denote the total length and number of double bonds of both acyl chains respectively. For example PC 34:1 is a phosphatidylcholine having total 34 carbons and contains one double bond.

This method of nomenclature has an exception of PE plasmogen species which are denoted by letter p in the name and in this case the acyl chain at the sn-1 position is replaced by an alkenyl group. SM and ST species are numbered according to the length and number of double bonds in the acyl chain attached to the sphingosine base and the hydroxylated STs are denoted by additional OH in their nomenclature. For example, ST 24:1 is a sulfatide with total length of 24 carbons and one double bond in the acyl chain attached to the sphingosine base, while its hydroxylated analoge is denoted as ST 24:1 (OH) [76]. Gangliosides contain one or more negatively charged sialic acid residues. They play important role in brain development, neuritogenesis, memory formation, synaptic transmission and aging [76; 84].

1.4.2 Analysis of Lipids

The zwitter ionic lipid species like PCs, SMs and PEs are easily detectable while PEs are less sensitively detected by MALDI-MS [85]. The PCs and SMs both have the choline head group which leads to high yield of positve ions [86]. The lipid species PS and phosphatidic acids (PAs) are also shown to be detectable with low sensitivity [6].
The phospholipids like PCs and SM that contain quarternary ammonium groups may prevent the detection of further species. This results in the difficult to detect other lipid species in the MALDI-MS screening of the crude biological mixtures that contain high amounts of PCs [85]. When small amounts of acidic phosphopholipids (PG, PS and PI) are to be detected in presence of high amounts of PCs or SMs, it is advantaceous to use negative ion mode [87]. The strongly acidic PAs contain two negative charges at physiological pH (7.4) and thus requires three cations to produce a single positive charged ion whereas addition of only one cation is required for the detection as negative ion [87]. The sphingolipids can be analysed as intact molecules and they generate several peaks due to the heterogeneity in the sphingosine or acyl chains [87].

Lipid analytes fragment uncontrollably that results into loss of specificity and sensitivity [88]. For example, gangliosides are made up of a ceramide backbone with sialyated oligosaccharides attached, lose sialic acid residues [6]. Thus, the matrices used for lipids should be different than those used for proteins. For high resolution imaging of sphingomyelins (SM) and phosphatidylcholines (PC) a mixture of dihydroxyacetophenone (DHA), heptafluorobutyric acid (HFBA) and ammonium sulphate can be used [87]. For sulfatides and phospholipids 9-aminoacridine was shown to be suitable [89].

Sphingolipids

Sulfatide
(negative charge)

Ganglioside
GM1 (18:0/d18:1)

Glycerophospholipids

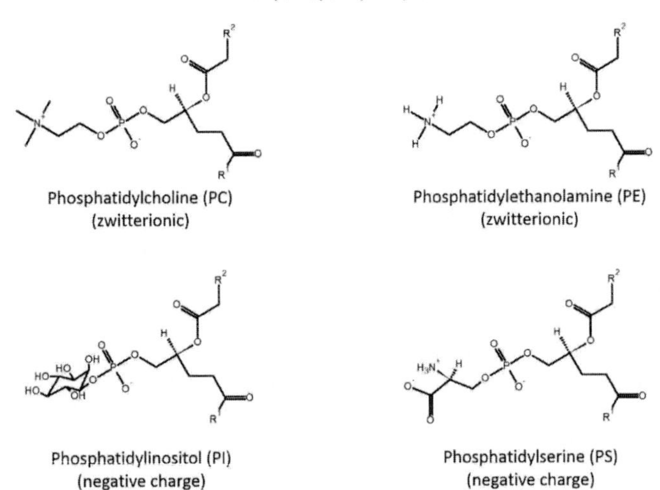

Figure 1.4: Structures of different types of brain lipids. R^1 and R^2 represent the chains of fatty acids. Adapted from [76]

1.4.3 Developing New Matrices

There are large number of matrices available for different analyte classes and for different modes, positive or negative, of mass spectrometry. And one might think that the field of matrix development has reached to its peak and there is not need to develop new matrix any more. But that is not the case. Each matrix which is used today has one or more limitations and/or drawbacks and the matrix might need to be replaced (see Table 1.2).

For example DHA [90] and DHB [91] are the most commonly used matrices for lipids in negative ion mode. DHB tends to form large crystals which causes molecular delocalisation and also poor spot-to-spot reproducibility whereas DHA sublimates under high vacuum and can be used once for short time periods for detection [92; 93]. para-nitroaniline (PNA) also exhibits the drawback of high vapour pressure similar to DHA [94]. 2,4,6-trihydroxyacetophenone (THAP) is a recommended matrix for various lipid classes like neutral storage lipids (triglycerols), polar membrane lipids (phospholipids, sphingolipids and glycosphingolipids), however, it forms large crystals [95]. 2-mercaptobenzothiazole (MBT) is a well-suited matrix for proteins [95] and lipids [96] because it can offer a good S/N ratios, and high spot-to-spot reproducibility, thanks to its small crystals and homogeneous crystallisation. But it also has limitation that the number of well-defined peaks for lipids in negative mode between *m/z* 780 and 880 is limited [95] (Table 1.2. 9-AA is often used for phospholipid and sulfatide detection in negative ion mode but it has unfavourable toxicity profile, high variability in automated data acquisition and it has an absorption minimum at the laser irradiation wavelength 355 nm, thereby leading to high ionisation energy threshold [97].

Matrix	Limitations/Drawbacks
2,5-Dihydroxybenzoic acid (DHB)	Large crystals, molecular delocalisation, poor reproducibility
Dihydroxyacetophenone (DHA)	High vapour pressure, sublimates under high vacuum
p-Nitroaniline	High vapour pressure
Trihydroxyacetophenone (THAP)	Large crystals
9-Aminoacridine (9-AA)	high variability in automated data acquisition, toxic
2-Mercaptobenzothiazole	All lipids between 780 Da and 880 Da cannot be detected

Table 1.2: Commonly used MALDI matrices and their limitations

1.5 Outline of the Thesis

This thesis is made up of three main chapters *viz.* literature survey, results and discussion, and the experimental section. The results and discussion chapter is further divided into sub-sections. At the beginning, the synthesis of different matrices are rationalised and the design and synthesis of those matrices is explained in detail. In the next subsection, new matrices for different analyte classes are introduced along with the set of the experimental conditions used. In the following subsection, statistical analysis for the matrices is explained for selected matrices. The experimental section consists of general information about the materials and chemicals, synthesis, sample and matrix preparation, X-ray crystal structure data, statistical analysis and different techniques (NMR, UV-Vis, MALDI-MS) used throughout this work. The later part of this chapter is made up of characterization of all the compounds synthesised.

2

Literature Overview

2.1 Introduction

Matrix assisted laser desorption/ionisation (MALDI) is one the most commonly used ionisation techniques for the analysis of biomolecules by mass spectrometry. The choice of matrix can have a large effect on the peak intensities, relative abundance of the analyte, fragmentation of the analyte and also the types of detectable analyte ions. Considerable research has been conducted to define what makes a suitable matrix for MALDI.

In general, however, the only persistent rule is that the matrix should absorb the laser energy at the laser wavelength, such that excessive internal energy is not imparted to analyte but at the same time it should be high enough to promote the desorption/ionisation process. The desirable qualities of a good matrix are vacuum stability, chemical inertness, ability to generate a homogeneous matrix-analyte mixture. In reality, most of the matrices cannot meet all of these requirements, so depending on applications one or the other compromises are made.

There are matrices that are successfully used for the MALDI analysis like 4-hydroxy-α-cyanocinnamic acid (HCCA) and 2,5-Dihydroxybenzoic acid (DHB). Recently, 4-chloro-α-cyanocinnamic acid (ClCCA) has been introduced as a new rationally designed MALDI matrix. It has been shown to have a great potential for peptide analysis leading to an improvement in the sequence coverage and sensitivity compared to HCCA. This has been hypothesised to be due to lower proton affinity of ClCCA than that of HCCA.

MALDI-MS has become a powerful tool for detection, identification and characterisation of peptides, proteins and lipids from complex mixtures. More recently, MALDI-imaging was initiated and now is developed in several laboratories. These developments were applied to several biological questions demonstrating the potential of MALDI-MS in the medical field.

However the direct analysis of crude samples require maximum resolution, sensitivity and dynamic mass range since tissues are very complex moieties presenting a wide range of molecules interacting together, some of which are expressed in very low concentrations compared to some others or presenting very close m/z values. To obtain optimal performances MALDI needs new developments either in sample preparations or in instrumentation, for example, new detectors and SMALDI for high resolution imaging. One alternative is to develop new matrices as matrix is fundamental in desorption/ionisation process, which contributes to the spectral quality i.e. peak resolution, sensitivity, intensity and signal-to-noise ratio. Actually, HCCA, ClCCA, DHB and sinapinic acid (SA) are the matrices commonly used for peptide and protein analysis. For direct analysis of tissues the choice of matrix is even more instrumental as homogeneity of crystallisation becomes crucial. HCCA, ClCCA and SA are good candidates while DHB crystalisation is too heterogeneous except in case of special spotting systems are used. Direct tissue analysis leads to a lowered spectral quality. Therefore, development of new matrices for tissue analysis is a need for addressing such problems.

The qualities of a new matrix compared to HCCA, DHB, SA, 9-AA and ClCCA must be a

- better spectral quality in terms of resolution, intensity, noise, sensitivity, number of compounds detected and identified and tolerance for contaminants and complexity of the sample.

- better crystallisation on tissues, covering capacity, homogeneity of crystallisation and crystal sizes and time needed for crystallisation

- better vacuum stability

- better resistance to high frequency lasers

Although it has been claimed that the performance of the matrix depends largely on the chemical properties, this aspect is less studied and this aspect is still under discussion [51]. One of the limitations for the systematic study of the structure of matrix and its performance being the reliance on the commercial availability of the compounds. One approach to overcome this limitation would be an intelligent synthesis of the potential matrices which would provide a vast range of functionalities to study the structure-activity relationships [98].

2.2 Synthesis of Matrices

2.2.1 HCCA as a Basic Scaffold for the Synthesis of New Matrices

HCCA is the most commonly used matrix for MALDI-MS for the analysis of proteins and peptides. Many of its analogues like Cl-CCA, sinnapinic acid, PCCA, 2,4-difluoro-α-cyanocinnamic acid (Di-FCCA) [96], (*E*)-2-cyano-3-(naphthalen-2-yl)acrylic acid (NpCCA) and (2*E*)-3-(anthracen-9-yl)-2-cyanoprop-2-enoic acid (AnCCA) [99] have been developed for protein and/or lipid detection. There are few successful attempts to develop matrices using HCCA as basic scaffold [96; 99; 100]. Amino acids, even proteins and other large molecules can be ionized more easily with UV light, when they were mixed with conjugated aromatic compounds like cinnamic acid derivatives, because then, they had a wide absorption band in the UV range [20].

2.2.2 Synthetic Ways To *E*-Cinnamic Acid Derivatives

2.2.2.1 The Perkin Condensation Reaction

The reaction between aromatic aldehydes and acetic anhydride in the presence of potassium or sodium salt of the corresponding acid is an easy way to synthesise cinnamic acids (see Figure 2.2. Many other bases can catalyze the reaction, like carbonates, phosphates, sulfides, organic amines [101; 102].

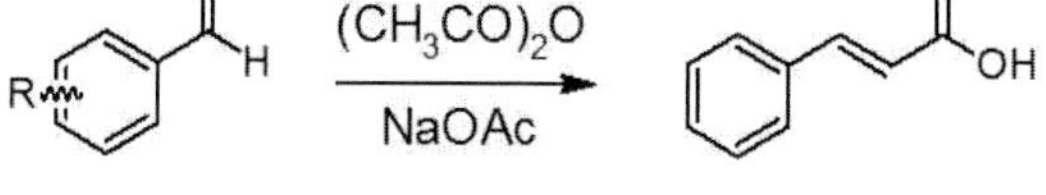

O H R $(CH_3CO)_2O$ NaOAc O OH

Figure 2.1: Scheme of the Perkin Condensation

The modified version of Perkin reaction can be used for the synthesis of α-substituted cinnamic acids where acetic anhydride is the solvent and a reactant at the same time, reacting with acetic acid derivatives (for example phenylacetic acid, see Figure 2.2) and usually amine base is applied instead of inorganic salts. [103].

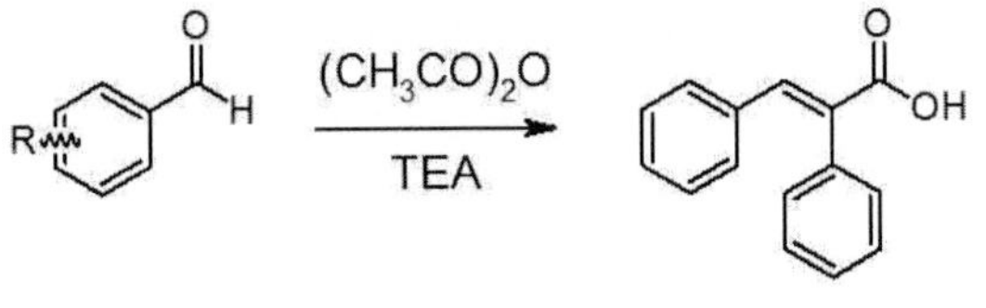

$$\text{R-C}_6\text{H}_4\text{CHO} \xrightarrow[\text{TEA}]{(CH_3CO)_2O} \text{PhCH=C(Ph)COOH}$$

Figure 2.2: Scheme of the modified Perkin Condensation for the synthesis of *E*-2-phenylcinnamic acid

The main disadvantage of the Perkin reaction is the formation of resin-like side products for heterocyclic aldehydes, and the low yield, when the aromatic aldehyde contains one or more electron donor groups.

2.2.2.2 The Knoevenagel Condensation

One of the most effective synthetic methods, which can lead to cinnamic acids is the Knoevenagel condensation [104]. The reaction between aromatic aldehydes and acetic anhydride in presence of a base via decarboxylation gives cinnamic acid as a product. The reaction scheme is shown in the Figure 2.3

$$RCHO + CH_2(COOR^1)_2 \longrightarrow RCH{=}(COOR^1)_2 \longrightarrow RCH{=}CHCOOR^1$$

Figure 2.3: Scheme of the Knoevenagel Condensation

Knoevenagel-Döbner modification

This modification works with aromatic aldehyde e.g. benzaldehyde and the esters e.g. diethylmalonate (see Figure 2.4) that can lose one α hydrogen and form carbanion in presence of piperidine (base) yielding diethyl-benzylidene malonate. The alkaline hydrolysis of the ester formed followed by acidification results into cinnamic acid.

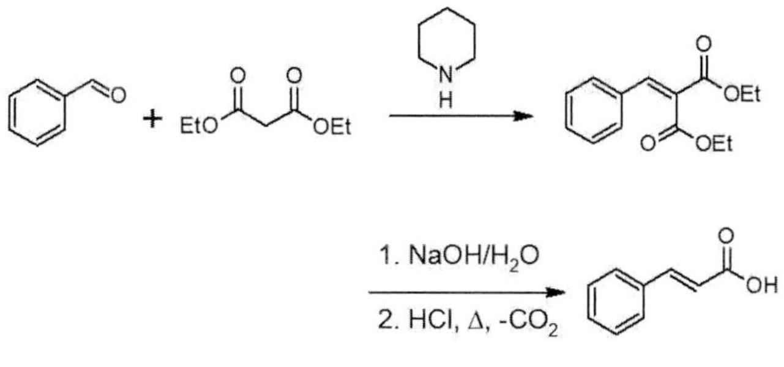

Figure 2.4: Scheme of the Knövenagel-Döbner modification

The main disadvantage is the problematic heat treatment for the last step of decarboxylation that needs 135–150°C for complete decarboxylation [105].

The Döbner condensation can also used with malonic acid and pyridine that results into about 90–100% yield most of the times and the hydrolysis and decarboxylation also work smoothly and in pot.

2.2.2.3 The Heck Coupling

The Heck coupling reaction (or Heck-Mizoroki reaction) [106; 107] takes place between alkyl or aryl halides and olefins in the presence of palladium-organophosphorous complexes and a base. The alkene has to contain at least one hydrogen and it performs better, if it contains EWG, e.g. esters like methyl acrylate, yielding cinnamoyl esters.

Using Pd/C as a catalyst for example, the Heck coupling reaction between iodobenzene and methyl acrylate in *N*-methylpyrolidine (NMP) as solvent and Pd/C as the catalyst under ultrasonic conditions (see Figure 2.5) yield methylester of cinnamic acid [102; 108].

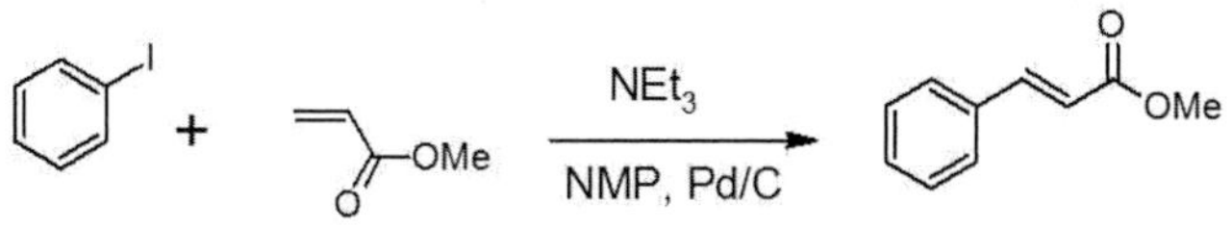

Figure 2.5: Scheme of the Heck coupling reaction between iodobenzene and methyl acrylate in NMP

Using $PdCl_2$ as a catalyst Cinnamic acid esters can be prepared (when X = COOMe)

from different aryl halides by using $PdCl_2$ as a catalyst under ultrasonic condition [102; 109]. Tetra butyl ammonium bromide (TBAB) is used as phase transfer catalyst while Na_2CO_3 acts as a base.

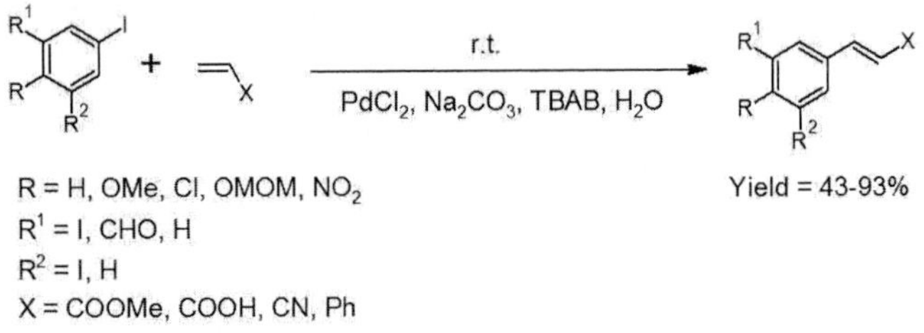

r.t.

$PdCl_2$, Na_2CO_3, TBAB, H_2O

R = H, OMe, Cl, OMOM, NO_2
R^1 = I, CHO, H
R^2 = I, H
X = COOMe, COOH, CN, Ph

Yield = 43-93%

Figure 2.6: Scheme of the Heck coupling reaction catalysed by $PdCl_2$ (from [102]

2.2.3 Synthetic Ways to *Z*-Cinnamic Acid Derivatives

2.2.3.1 Photoinduced Isomerisation of *E*-Cinnamic Acids

Molecules with conjugated electronic system can be excited with UV photons. The excited molecule becomes very reactive, but it can be relaxed in nonchemical ways like visible photon emission (fluorescence or phosphorescence) or trigger thermal reactions. However, the quantum efficiency is only around 0.2 for most of the organic reactions. On the contrary, in most cases, the excited molecule starts chemical reactions like rearrangement by isomerization, which is a useful tool to produce Z-cinnamic acids from the more easily available (due to structural stability) *E*-isomers. The isomerization of *E*-cinnamic acid takes place without bond cleavage via a one-bond flip around the double bond (diabatic photochemical reactions),(rotational barrier = 50−60 kcal/mol range) thus, the UV light (e.g. at 254 nm) has enough energy to induce the isomerization (112 kcal/mol)[110].

Both *E* and *Z* isomers have a wide absorption band in the UV range; thus, when they are irradiated with a mercury lamp (254 nm emission maximum) an $S0 \rightarrow S1$ transition takes place, where the excited molecule can freely rotate around the double bond. The two main ways possible to return to the ground state are: the fluorescent way and the non-radiative relaxation. The probability of the different types of relaxations is not equal; thus, one gets different isomer distribution: for *E*-cinnamic acid, 75:25, *Z*:*E* ratio was found after 72 h irradiation at 254 nm [110; 111].

The choice of solvent is critical in photochemistry as well. There are two different types, the reactive (water, alcohols, aromatics, alkyl halogenides), absorbing UV light and the unreactive (alkanes, esters), which do not absorb in the UV range. For photoisomerization, the use of unreactive solvents is recommended.

2.2.3.2 *Z*-selective Catalytic Hydrogenation of Alkynes

A very common method for the preparation of *Z*-cinnamic acid is the hydrogenation of phenylpropiolic acid derivatives with Lindlar's catalyst (Figure 2.7), which is a heterogeneous Pd catalyst poisoned with lead on the surface of $CaCO_3$ or $BaSO_4$. The syn-addition of hydrogen to the triple bond leads to the *Z*-isomer [112]. This method results in to good yields and *Z*:*E* selectivity was found to be 95:5 [112].

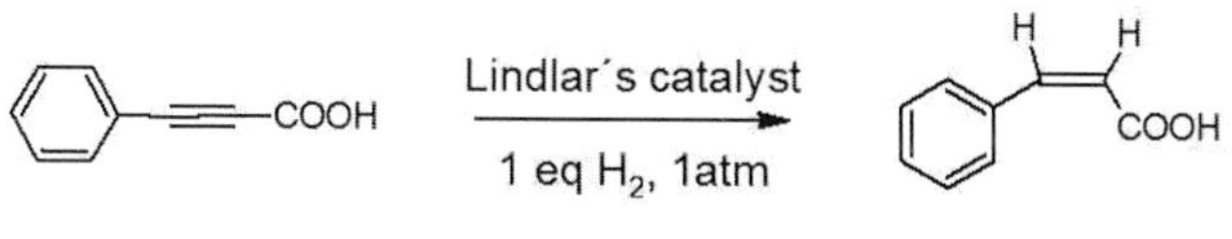

COOH — Lindlar´s catalyst, 1 eq H_2, 1atm → H H COOH

Figure 2.7: Scheme of the hydrogenation reaction using Lindlar's catalyst (from [102])

2.2.3.3 The Horner-Wadsworth-Emmons (HWE) Olefination

The Horner-Wadsworth-Emmons (HWE) [113; 114; 115] method is one of the most favorable techniques for the stereoselective synthesis of olefinic esters. Its mecha-nism is similar to the Wittig olefination, except that the reagents are phosphonates instead of phosphoranes, and the alkyl group should contain an EWG, e.g. ester or nitrile groups. The different phosphonates can be synthesized utilizing the well-known Michaelis-Arbuzov rearrangement [116; 117], where an alkyl halide reacts with organic phosphites to give the desired phosphonate esters. If the phosphonate has large side group syn addition becomes sterically unfavourable, due to steric hindrance between the phosphonate and the ring of the aldehyde, and this will facilitate the anti- route, which leads to the *Z* isomer.

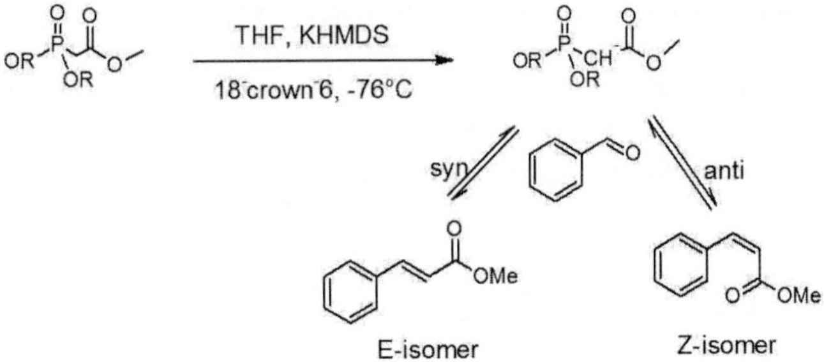

Figure 2.8: Scheme of the Horner-Wadsworth-Emmons (HWE)

2.3 Multivariate Analysis

The objective of this section is to give a general introduction to and to discuss the multivariate data analytical technique considered in this thesis, principal component analysis (PCA). This method is appropriate for analysing high dimensional data sets such as mass spectra where the objectives are to obtain an overview of the data. PCA is a multivariate projection method that is designed to extract the variation in a multivariate data set **X**.

The term chemometrics was first introduced in 1971 to describe the application of mathematical, multivariate statistical and other logic-based techniques in the field of chemistry, in particular analytical chemistry. The application of chemometrics has found considerable success in three areas, (a) calibration and validation of biological measurements (multivariate calibration); (b) optimisation of chemical measurements and experimental procedures; and (c) extraction of chemical information from analytical data (classification, pattern recognition, clustering)[118].

2.3.1 Principal Component Analysis

2.3.1.1 Theory of Principal Component Analysis

Principal component analysis (PCA) is the method used by chemometricians for data compression, information extraction and preliminary visualisation of observations or samples [119; 120]. The main function of PCA is to reduce the high-dimensionality of the multivariate data to a few dimensions that capture the main source of variability in the data. The new space is defined in terms of principal components (PCs) that are a linear combination of the original variables.

The weights of the individual variables in the principal components are termed loadings. They are useful for identifying the important variables in individual PCs, and also contain information on how the variables relate to each other. Scores are the coordinates of the original data in the new space and contain information on how samples relate to each other with groups of samples indicating similar behaviour [121].

2.3.1.2 Algorithm of PCA

There are a number of PCA algorithms, *viz.* non-iterative partial least squares (NI-PALS), the power method (POWER), singular value decomposition (SVD) and eigen-value decomposition (EVD). The EVD algorithm is briefly described in this section. According to this algorithm, PCA is based on the eigenvalue decomposition of the covariance or correlation matrix of the original data [122].

For a given data matrix **X** with m rows and n columns, the covariance matrix of is defined as:

$$\mathrm{cov}(\mathbf{X}) = \frac{\mathbf{X}^T\mathbf{X}}{m-1} \tag{2.1}$$

provided that the columns of **X** have been 'mean centre'. If the columns of **X** have been 'autoscale', equation 2.1 gives the correlation matrix of **X**. PCA decomposes the data matrix **X** as the sum of the outer product of vectors t_i and p_i:

$$\mathbf{X} = t_1 p_1^T + t_2 p_2^T + \ldots t_M p_M^T \tag{2.2}$$

In equation 2.2, the t_i vectors are known as score vectors while the p_i vectors are known as the loading vectors. Equation 2.2 can be written in the following matrix form:

$$\mathbf{X} = T\,P^T \tag{2.3}$$

where $T = [t_1\ t_2\ \ldots\ t_m]^T$ is the score matrix and $P = [p1\ p2\ \ldots\ pm]^T$ is the loading matrix. In the PCA decomposition, the p_i vectors are the eigenvectors of the covariance matrix:

$$\mathrm{cov}(\mathbf{X})p_i = \lambda_i p_i \tag{2.4}$$

where λ_i is the eigenvalue associated with the eigenvector p_i.

The PCs are arranged in descending order based on the eigenvalues: $\lambda_1 > \lambda_2 > \ldots > \lambda_m$ i.e. the first PC explains the greatest amount of variability with the second PC explaining the next greater amount variability in **X**. As many PCs as variables, if m > n, can be calculated but the majority of the variability will be captured in the first few PCs.

Therefore the PCA decomposition of **X**, can be represented as:

$$\mathbf{X} = t_1 p_1^T + t_2 p_2^T + \ldots t_k p_k^T + \ldots + E \tag{2.5}$$

where E is the residual matrix. In practical applications K must be less than or equal to the smaller dimension of X, i.e. $k \leq \min\{m, n\}$. Since E typically contains noise, it has the effect of noise filtering and will not cause any significant loss of useful information.

In this thesis PCA was used to the parameters relating structural features of matrices like molecular weight, logP values etc and performance parameters like relative mean S/N ratios.

2.3.1.3 Interpretation of PCA

The loading plots of the PCA can be plotted for different combinations of PCs with significant variance. Each vector in the loadin plot represents different parameter used for PCA. The direction of the vector explains about the correlations between several parameters. The vectors in same direction are said to be positively correlated while those in the opposite directions are negatively correlated. If the vectors are orthogonal then they are said to be not correlated to each other. The smaller the angle between these vectors greater is the positive correlation between them. The length of the vectors tells how good that parameter is loaded on the PCA. In other words, it tells about its influence on the loading.

Score plots takes in to account all the data sets that have considered for the PCA, in this case different matrices. Similar data sets i.e. matrices would appear close to each other depending on the parameters taken into account. Therefore, similar matrices would group together.

Scree plot explains how many PCs are significant and should be considered for better interpretation of PCA.

3

Goals and Objectives

Since its discovery, continuous efforts are being put into uncovering the working principles of MALDI-MS. It is an undisputable fact that the matrix plays a prime role in the laser desorption/ionisation process of the anlayte molecules. The process involves the matrix absorbsing laser energy and transfering some part of it to the analyte molecules such that the molecules are desorbed from the surface. As a result different sample preparation protocols have been developed to improve reproducibility and sensitivity of the mass spectrometric evaluations. Research till date documents several models that elucidate the primary and secondary ionisation processes. Chapter 1 portrays all these features of the MALDI-MS in detail.

As a part of scientific achievement in research on MALDI-MS, a large number of matrices have been developed for different analyte classes. Unfortunately, the process for identification of novel matrices is not well established. The development of new matrices is still a matter of trial and error. Research shows that a very few matrices like p-chloro- α-cyanocinnamic acid (ClCCA) have been rationally developed.

Considering the facts presented so far, the main goal of this doctoral work was to develop novel matrices for different analyte classes for positive or negative ionisation mode of mass spectrometry. The aim was to discover new matrices that would qualify as better than the matrices currently in use in many aspects. Besides being easy to synthesise, the new matrix should result in superior performance in terms of S/N ratios, reproducibility, sensitivity and selectivity.

To further explore the aspects of matrices, an important goal was to explore if here exists a relationship between the structure of matrix and its performance as matrix. Taking into consideration the relationship aspects, many questions remain unanswered especially which physical and chemical properties are essential to be a matrix? Which structural features in the matrix are important to enhance its performance? Does geometry play any role in deciding matrix performance? Specific goals were set to address these questions.

To achieve the above mentioned research goals, a compound library of cinnamic acid was designed and approximately 200 compounds were synthesised. These potential matrices were used as matrices for Bovine Serum Albumin (BSA) digest and brain total lipid extract (BTLE) detection using 337 nm laser. Out of these compounds 59 structurally related matrices were evaluated separately for sulfatides from BTLE using two different lasers (337 and 355 nm). Details of the synthesis and results can be found in Chapter 4.

4

Results and Discussion

Goals of the chapter:

- To putforth the results obtained by different types of matrices for the detection of peptides fron BSA digest and lipids from BTLE, protein calibration standard mixtures and compare them with the standard matrices
- To putforth the results obtained by using phenylcinnamic acid derivatives as MALDI matrices for lipids, especially sulfatides, in negative ionisation mode of mass spectrometry.
- To compare the results obtained by using phenylcinnamic acid derivatives for two different lasers with different laser wavelengths
- To try to establish a relationship between matrix structure and its performance, if possible.

In the earlier part of this chapter, how the new matrices are generally developed and what are the challenges in the development process will be discussed. It will be followed by the discussion about design and synthesis of new potential matrices. Then the results obtained by few of the best matrices from each type of HCCA derivatives and their comparison with the results obtained using standard matrix will be shown.

The major part of this chapter will be dedicated to the most successful matrix p-phenyl-α-cyanocinnamic acid amide (p-PhCCAA) and its analogs, that is, biphenyl derivatives. The observations and results obtained in the attempt to discover the structural factors of matrix that affect its performance will be discussed in detail.

4.1 Development of New MALDI Matrices

For both positive and negative mode of MALDI-MS, several matrices are successfully developed as well as commercially available for different analyte classes like proteins, peptides, lipids and oligosaccharides etc. For example HCCA, ClCCA and DHB are the most commonly used matrices for the detection of proteins and peptides in positive ion mode while 9-AA and DHB are used for lipids in negative ion mode. Several ionic liquids and solids are used as matrix for the detection of glycoproteins, oligosaccharides, biodegradable polymers.

The choice of MALDI matrix is a key element in the MALDI-MS measurements and yet very few facts about it are scientifically established. One of the most detailed studied fact is that the matrix is a small organic compound that can absorb laser energy at the laser wavelength used and transfers this energy to the analyte molecule.

4.1.1 Factors Affecting the Quality of MALDI Spectrum

The development of a MALDI matrix by trial and error method is not at all a great way to choose a suitable matrix for the detection. The identification process of novel matrix is not well established until now. It might be due to the fact that a large number of factors play a significant role in improving or destroying the quality of MALDI spectrum. Following is a brief account of different experimental, instrumental and matrix parameters that have influence on the spectrum.

4.1.1.1 Instrumental Factors

- the type of laser used, laser focus, pulse duration, beam modulation and laser energy
- the target surface structure

4.1.1.2 Experimental Factors

- sample preparation
- method for drying i.e the time given to allow crystallisation
- whether or not buffer is used
- pH value of the system
- solvent-matrix ratio
- solvents used for matrix and analyte solutions and their concentrations

4.1.1.3 Matrix Properties

- hydrophobicity (logP or logD)
- UV-Vis absorbtion properties
- vacuum stability

Fundamental matrix properties for the matrix assisted ionization process can be determined experimentally, but due the large number of influential parameters for matrix performance the pathway leading to the discovery of new MALDI-MS matrices is not well defined. To cover this gap it is obvious that rational design and evaluation strategies used in medicinal chemistry should be transferred to MALDI matrix development. One strategy to achieve this goal is to identify necessary general properties of MALDI matrices and to use these properties as constraints for the virtual screening of compound libraries to improve the hit likelihood for the identification of novel matrices.

4.2 Design and Synthesis of Compound Library

HCCA is the most commonly used matrix and it is the matrix which has been used most of the times to synthesise analogs for the development of new matrix for MALDI-MS. For example, PCCA, SA, gentisic acid all are HCCA analogs. The HCCA scaffold also have a potential of synthesising a large number of derivatives by substitution on the aromatic ring and also at the carboxy carbon. Knoevenagel condensation has a potential to synthesise cinnamic acid derivatives in a single step and was employed to design and synthesise a library of compounds related to

HCCA as shown in Figure 4.1 HCCA was considered as a basic scaffold and the library of its derivatives was synthesised and the compounds were treated as potential MALDI matrices.

The Figure 4.1 shows different types of compounds synthesised, e.g.the biphenyl derivatives, pyrrole derivatives, dienyl derivatives. All these compounds were tried as matrix for the detection of BSA digest in positive mode of ionisation and different lipid classes from the natural complex lipid extract i.e. brain total lipid extract in negative ionisation mode of mass spectrometry.

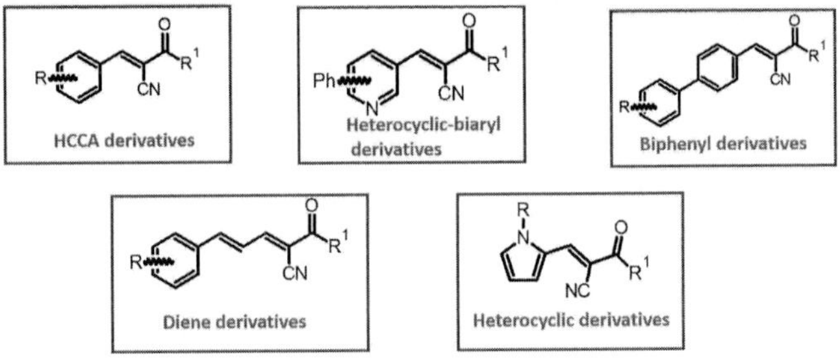

Figure 4.1: Different types of HCCA derivatives synthesised by Knoevenagel Condensation

4.3 The Overview of MALDI-MS Performance of Synthesised Compounds

As explained in the section 4.2 the HCCA matrix was considered as the basic structural scaffold for the compound library designing and then different types of HCCA derivatives were synthesised under Knoevenagel condensation conditions. All the synthesised derivatives were tried as matrices for the detection of peptides from BSA digest and lipids from brain total lipid extract (BTLE) in both positive and negative ion mode of mass spectrometry for 337 nm N_2 laser. And some of the matrices were tested using 355 nm Nd:YAG laser for lipids in negative ion mode.

4.3.1 Phenylcinnamic Acid Derivatives

A phenylcinnamic acid derivative p-PhCCAA was synthesised (by Miss Martina Porada, University of Applied Sciences, Aalen) as a potential matrix. It exhibited outstanding matrix qualities for lipid detection in negative ion mode.

This observation led to synthesis of more of its derivatives.

A compound library of total 59 phenylcinnamic acid derivatives was designed and synthesised (Figure 4.2). The matrices in the compound library are numbered in order of increasing logP values.

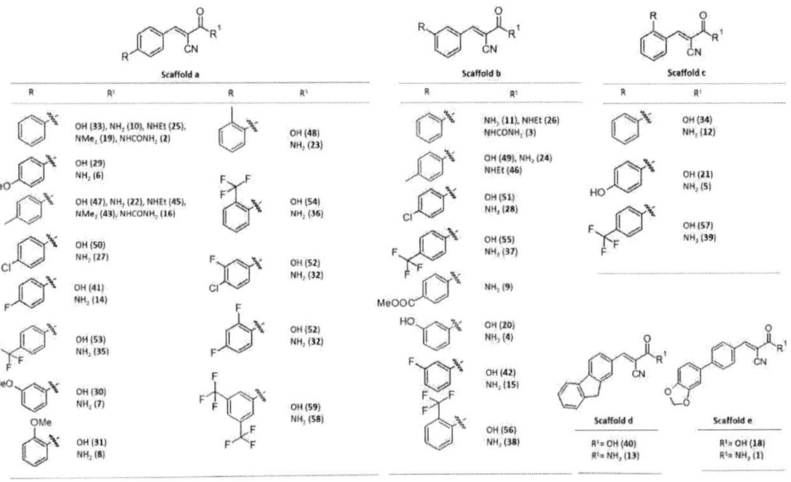

Figure 4.2: Structures of different scaffolds: scaffold a) para-phenyl-α-cyanocinnamic acid derivatives; scaffold b) meta-phenyl-α-cyanocinnamic acid derivatives; scaffold c) ortho-phenyl-α-cyanocinnamic acid derivatives; scaffold d) Fluorenyl-α-cyano-acrylic acid derivatives; scaffold e) 1,4-benzodioxolyl-α-cyanocinnamic acid derivatives.The indents show functional groups at carboxy position for the scaffolds. Different substituents on second ring are denoted by R^2, R^3 and so on. The matrix number is indicated in the brackets.

Structural diversity in the chemical library was introduced by modification of the carboxylic acid moiety and variation of the relative substitution pattern of both substituents at the first aromatic ring. Out of the 59 compounds, 32 p-biphenyl (Scaffold a), 17 m-biphenyl (Scaffold b), 6 o-biphenyl (Scaffold c), 2 fluorenyl (Scaffold d) and 2 1,4-benzodioxolyl (Scaffold e) derivatives were synthesized (Figure 4.2). To examine electronic and steric effects of small substituents various electron donating and electron withdrawing substituents on the second aromatic ring were incorporated like halogens, methyl, hydroxyl, methoxy, trifluoromethyl and ester groups. Additional diversity was introduced by replacing the carboxylic acid functionality by amide, ethylamide, dimethyl amide and acetylurea functionalities. Acid and amide derivatives were synthesized to examine the role of the acid/base chemistry for the performance of the matrices in the negative ion mode. *N*-alkyl, and *N,N*-dialkylamides were synthesized to examine the role of intra- and inter-residual hydrogen bond interactions of the amide functionalities. Acetylurea derivatives were synthesized to examine the role of the N-H pKa-values. The diversity of the synthesized chemical compound library is also reflected by its physical properties (e.g. logP, molecular mass, extinction).

Figure 4.3 explains the diversity of the compound library. The library design and synthesis is expalined in detail in the section 4.2. The library was designed and synthesised and then the properties like logP, pKa, Tanimoto scores were calculated and was found to be in the range within the Lipinski rules. The Tanimoto similarity score explained how diverse the compound library was. The Figure 4.3 shows 3-dimensional representation of diversity using parameters *viz.* logP, molecular weight and the Tanimoto score of the matrices. The Tanimoto similarity scores were calculated against matrix **10** as the reference structure. Different functional groups are represented by different colours and shapes of the spots. The spots overlap for some compounds as they are structural isomers, with same substituents attached at different positions on the phenyl ring and their calculated molecular weights, logP and Tanimoto scores are identical. The physical properties of the designed library of compounds represent a molecular mass range from 248 to 385 g/mol Similarly, the compounds exhibit high diversity in their absorption behaviour (λ_{max} = 257 to 349 nm; ε_{max} = 2288 to 45774 [L/(mol cm)]; ε_{355} = 0 to 28780 [L/(mol cm)]; ε_{337} = 39-33337 [L/(mol cm)]. The property values cover the following ranges: H-bond donor sites: 0-2; H-bond acceptor sites: 2-5; logP (hydrophobicity): 2.41-5.35

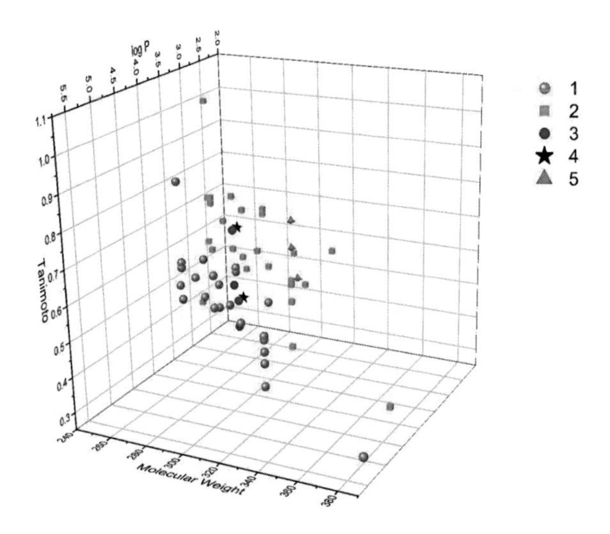

Figure 4.3: Diversity of compound library: For the diversity measurement, Tanimoto similarity score was calculated using online calculation tool ChemMine Tools. The colour and shapes of the spots are different for different functional groups (1: acid, 2: primary amide, 3: *N*-ethylamide, 4: *N,N*-dimethylamide, 5: acetylurea).

All the 59 matrices were evaluated with two different laser systems (337 and 355 nm wavelength) in negative ion mode for BTLE. The interesting results were obtained for p-Phenyl-α-cyanocinnamic acid amide (p-PhCCAA)(**10**), m-Phenyl-α-cyanocinnamic acid amide (m-PhCCAA)(**11**) and 4-Methyl-p-phenyl-α-cyanocinnamic acid amide (4-Me-p-PhCCAA)(**22**). Both the para derivatives p-PhCCAA (**10**) and 4-Me-p-PhCCAA (**22**) were the best performers for the 355 nm. The meta derivative m-PhCCAA (**11**) was excellent at 355 nm laser for sulfatide detection although its molar extinction coefficient is smaller at 355 nm.

The performance of all the 59 matrices were compared using two most common lasers i.e. 337 nm N_2 and 355 nm Nd:YAG lasers. The matrices are numbered according to their logP values. In other words, matrix **1** has smallest logP value and matrix **59** has the maximum one. The performance was compared with that of the standard matrix 9-AA (5 mg/ml) in (2-propanol/acetonitrile 2:3, v/v) as reference. The comparison is shown (Figure 4.4) for a representative sulfatide SM4s(40:1). The details of the results are discussed in the section 4.6.1.

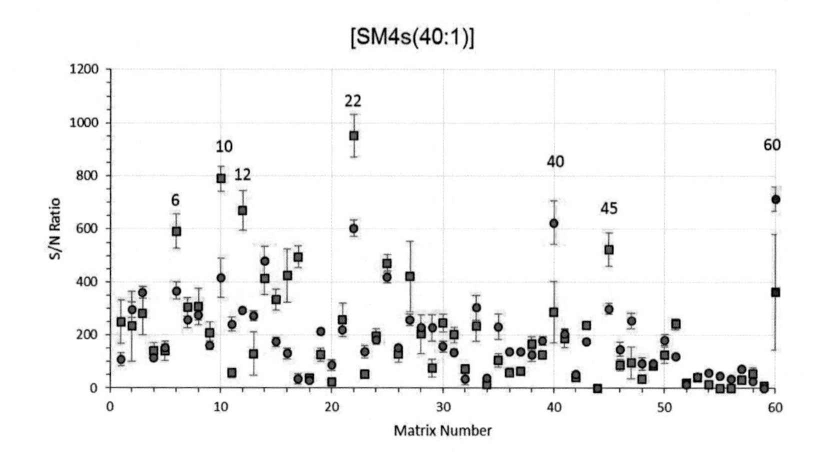

Figure 4.4: The performance of phenylcinnamic acid derivatives for the detection of a sulfatide SM4s(40:1) using 337 nm (blue circle) and 355 nm (green square) lasers in negative mode

The para derivatives were superior to 9-AA when used on 355 nm laser. The spectra comparing the new matrices p-PhCCAA (**10**) and 4-Me-p-PhCCAA (**22**) (5 mg/mL in ACN 50%) with the standard matrix 9-AA (**60**) (5 mg/mL in 3:2 IPA:ACN, v/v)for the BTLE (0.5 mg/mL in 3:2 methanol: chloroform, v/v) in negative ion mode of mass spectrometry are shown in Figure 4.5.

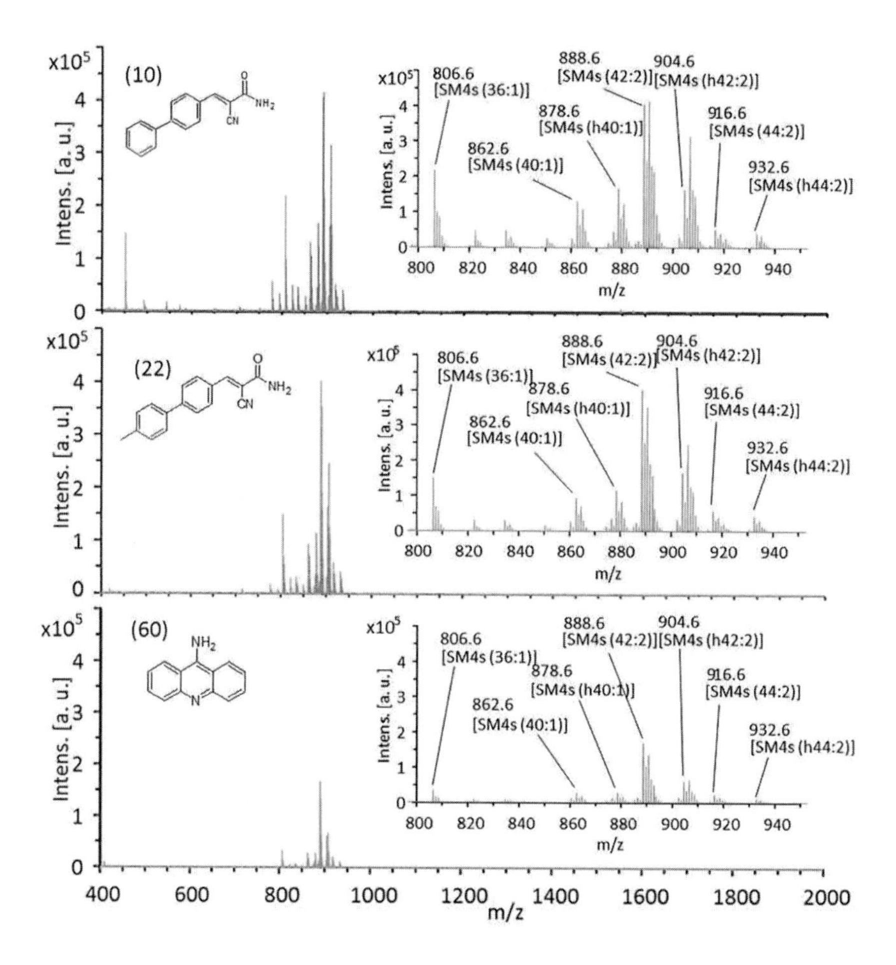

Figure 4.5: Mass spectra for the matrices numbered (**10**) and (**22**), compared to that of commonly used matrix 9-amino acridine (**60**) recorded in the negative ion mode using 355 nm laser

The Figure 4.6 compares the spectra generated by matrix **10**, **22** and 9-AA (**60**) for the detection of different gangliosides using 355 nm laser. Using 9-AA two GM1 gangliosides (36:1) and (38:1) could be detected [84]. Matrices **10** and **22** allowed the putative assignment of two more GM1 gangliosides Fuc-GM1 (36:1); (m/z = 1690.9

Da) and Fuc-GM1 (38:1); (m/z = 1718.95 Da) which are constituents of BTLE.

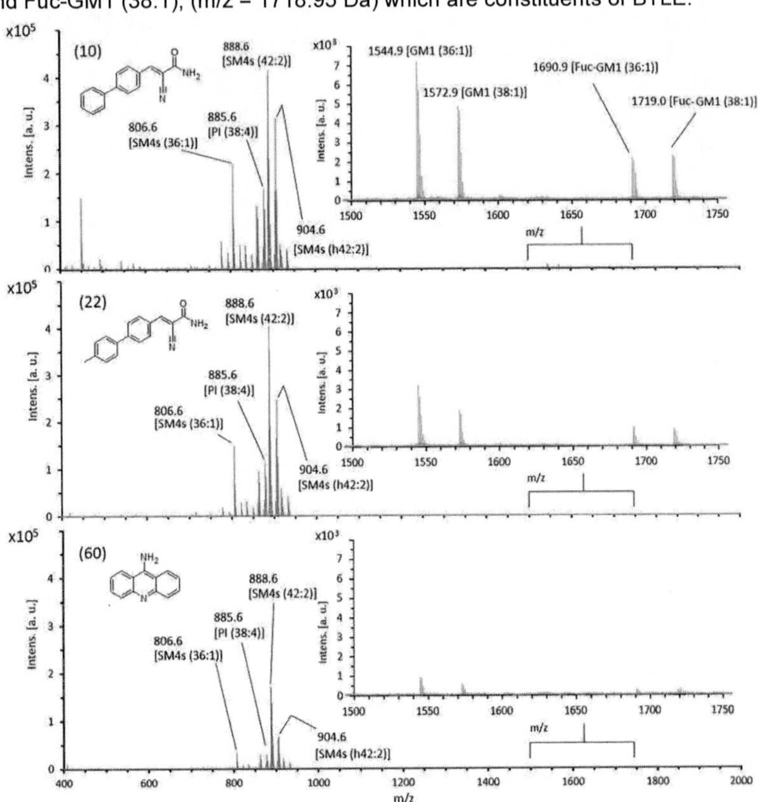

Figure 4.6: Mass spectra for the matrices numbered (**10**) and (**22**), compared to that of commonly used matrix 9-amino acridine (**60**) for gangliosides recorded in the negative ion mode using 355 nm laser

4.3.2 HCCA Derivatives

These derivatives were synthesised by Knoevenagel condensation between 2-cyanoacetic acid or 2-cyanoacetamide and corresponding benzaldehyde derivatives so as to obtain corresponding acid or amide derivative. All the structures can be found in the experimental section 6.1.2. The benzaldehyde derivatives having different electron donating groups like -OMe, -Me, -OH etc or electron withdrawing groups like CN, CF_3 at different positions on the aromatic ring of benzaldehyde were used for diversity in structural properties. Few HCCA deriavtives with hydrophobic alkoxy groups like isopropoxy, decyloxy were also synthesised to try for hydrophobic analytes. Total 42 HCCA derivatives were synthesised and tried as potential matrices for the detection of BSA digest (100 fmol/μL) and BTLE (0.5 mg/mL in 3:2 methanol: chloroform, v/v) in both positive and negative ion modes. Many of these derivatives act as ma-trix for both the analytes but their performances were not superior to the standard matrices. One candidate 4-hydroxy-3-ethoxy-α-cyanocinnamic acid (HECCA) must be mentioned here to exhibit excellent S/N ratios for BSA digest in positive ion mode using 337 nm laser (analysis done by Mr. Michael Schmich, University of Applied Sciences, Aalen). Future directions: The matrix must be optimised with different solvent mixtures, with different concentrations and different matrix-to-analyte ratios to obtain even better results, especially for the lipid detection in negative ion mode.

In Figure 4.7, the structure of a new matrix 4-hydroxy-3-ethoxy-α-cyanocinnamic acid (HECCA) (**61**) is shown. The Figure 4.8 exhibits S/N ratios obtained for different peptides from Bovine Serum Albumin (BSA) digest (100 fmol/μL) for HECCA in positive ion mode of mass spectrometry using 337 nm laser. The S/N ratios are compared with the commonly used matrix Cl-CCA. It is evident from the figure 4.8 that the S/N ratios for most of the peptides are better with the new matrix.

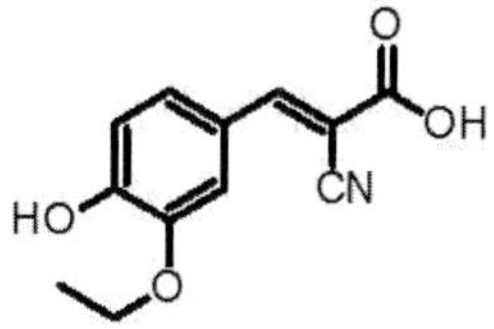

Figure 4.7: The structure of new matrix 4-hydroxy-3-ethoxy-α-cyanocinnamic acid (HECCA) (**61**)

4.3.3 Diene Derivatives

The replacement of the hydroxyl group in the conventional matrix 4-hydroxy-α-cyanocinnamic acid (HCCA) against the chlorine substituent in Cl-CCA leads to a hypsochromic shift. Consequently, increased performance is only achieved with MALDI-MS instruments with a nitrogen laser (337 nm) but not in instruments with a frequency-tripled Nd:YAG laser (355 nm) [123]. Thus, to achieve hypochromic shift in the UV-

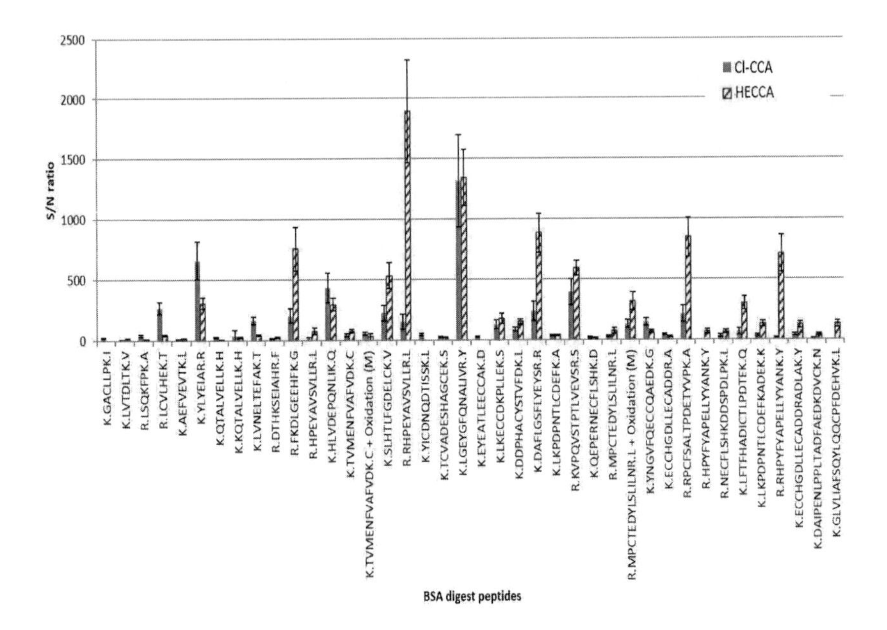

Figure 4.8: The comparison of S/N ratios for different peptides from BSA digest HECCA (**61**) and Cl-CCA in positive ion mode of MALDI-MS using 337 nm N_2 laser. The MALDI measeurements were carried out for 100 fmol/µL BSA digest and 5 mg/mL matrix solutions in 50% ACN

Vis absorbtion a compound library having an additional conjugation than in HCCA was designed and synthesied. Total 42 HCCA diene derivatives were synthesised. All the structures can be found in the experimental section 6.1.2.

All these diene derivatives were tested as MALDI matrices for the detection of BSA digest and BTLE in both positive and negative ion modes. Many of the derivatives were good matrices for the lipids in negative ion mode using 337 nm laser but further optimisation was necessary for better performance. Selected interesting matrices were optimised (by Miss Daniella Geissler, University of Applied Sceinces, Aalen) with different solvent mixtures and the two matrices 2-cyano-5,5-diphenyl-dienoic acid (DPDA) (**62**) and 2-cyano-5-phenyl-dienoic acid (PDA) (**63**) shown promising matrix behaviour for BSA digest in positive ion mode.

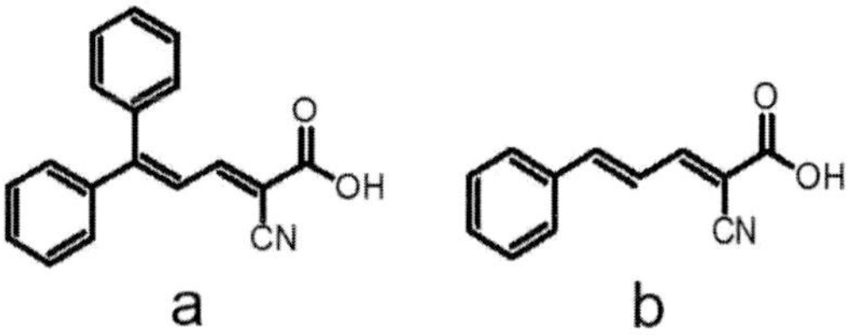

Figure 4.9: The structures of new matrices a) 2-cyano-5,5-diphenyl-dienoic acid (DPDA) (**62**) and b) 2-cyano-5-phenyl-dienoic acid (PDA) (**63**)

The structures and matrix performances are shown in Figures 4.9 4.10, 4.11.

Future directions: All the matrices should be optimised with different solvent mixtures and different matrix-analyte-ratios for BTLE. The matrices must be tried using 355 nm lasers as they were designed for the absorption of longer wavelength laser.

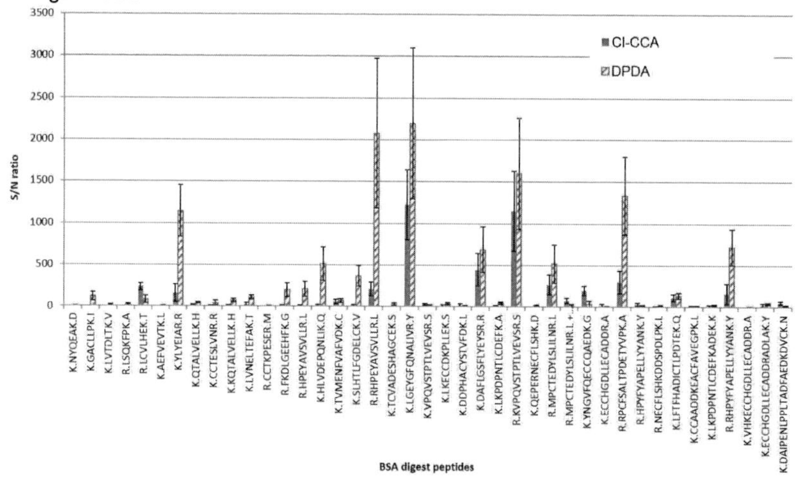

Figure 4.10: The comparison of S/N ratios for different peptides from BSA digest DPDA (**62**) and Cl-CCA in positive ion mode of MALDI-MS using 337 nm N_2 laser. The MALDI measurements were carried out for 100 fmol/µL BSA digest and 5mg/mL matrix solu-tions in 50% ACN and 0.5% TFA

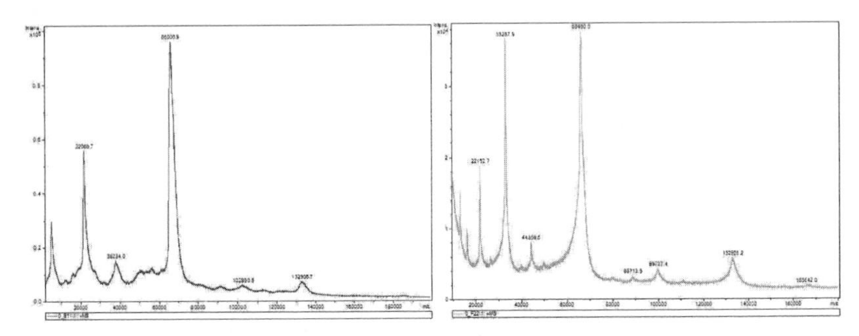

Figure 4.11: The comparison of spectra for different proteins from standard protein mix I using matrices PDA (**63**) and Cl-CCA in positive ion mode of MALDI-MS using 337 nm N_2 laser. The MALDI measurements were carried out for 100 fmol/µL BSA digest and 5mg/mL matrix solutions in 50% ACN and 0.5% TFA

4.3.4 Heterocyclic Derivatives

Different types of heterocyclic derivatives synthesized. Several of these derivatives were synthesised and complete list can be found in the experimental section 6.1.2.

The interesting results were exhibited by the pyrrole derivatives. These compounds displayed matrix behavior for the BSA digest (100 fmol/µL) in positive ionisation mode using 337 nm laser. The S/N ratios for 5 mg/mL in 1:1 (v/v) of Acetonitrile: H_2O solution of these matrices is shown in Figure 4.12

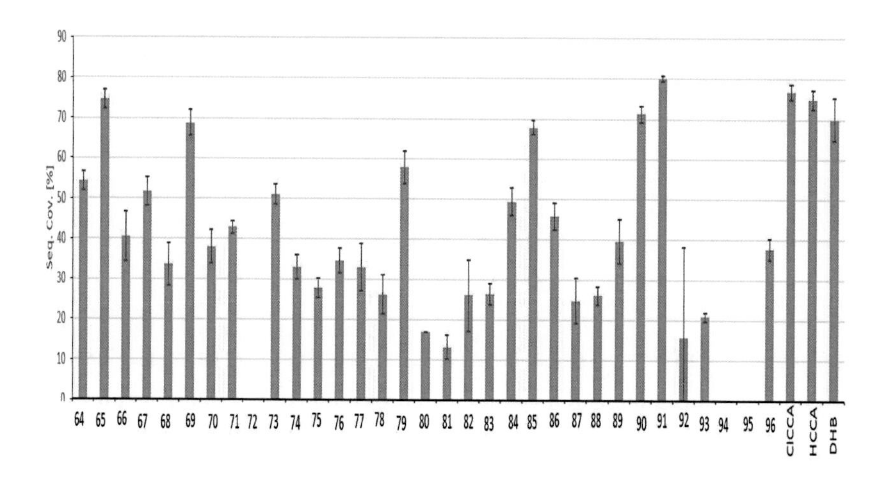

Figure 4.12: The comparison of sequence coverage in percentage for the pyrrole derivatives and standard matrices for 100 fmol/μL BSA digest and 5mg/mL matrix solutions in 50% ACN in positive ion mode using 337 nm laser

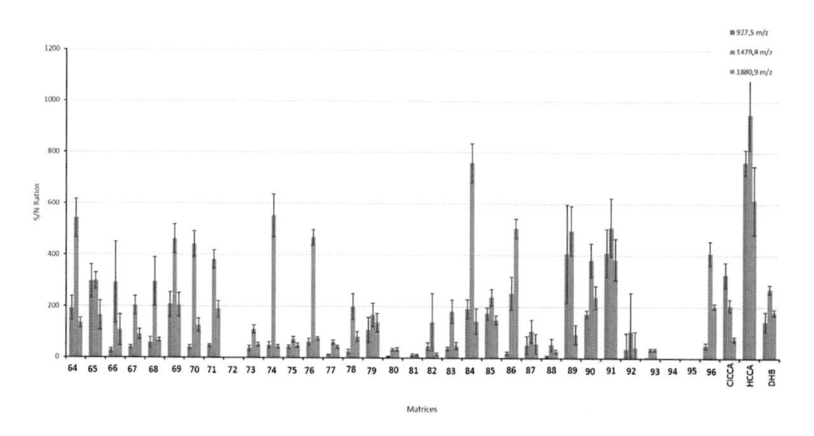

Figure 4.13: The comparison of S/N ratios for the pyrrole derivatives and standard matrices for 100 fmol/μL BSA digest and 5 mg/mL matrix solutions in 50% ACN in positive ion mode using 337 nm laser

Both the Figures 4.12 and 4.13 show how good the pyrrole derivatives are for the BSA digest in positive ion mode. The Figure 4.15 compares mass spectra of matrix (**90**) with standard matrix HCCA for BSA digest (100 fmol/µL) in positive ion mode (by Mr. Henning Blott, University of Applied Sciences, Aalen, Germany).

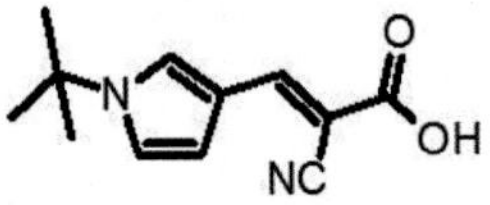

Figure 4.14: The structure of matrix **90**

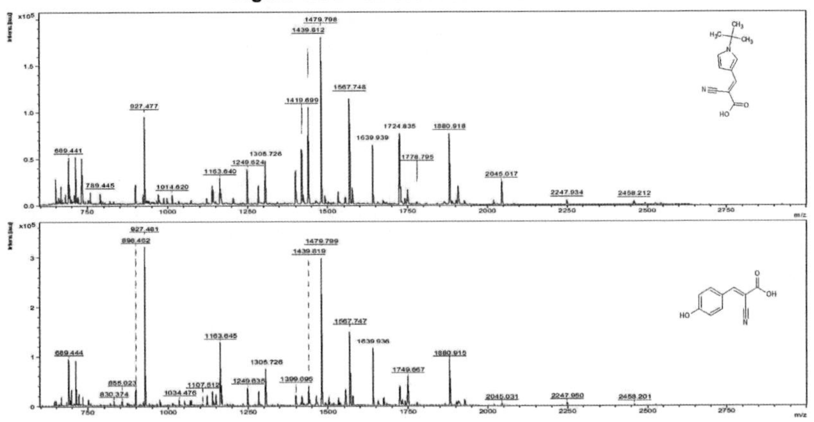

Figure 4.15: The comparison of mass spectra obtained using matrix **90** and HCCA, 5mg/mL solutions in 50% ACN for 100 fmol/µL BSA digest in positive ion mode using 337 nm laser

The pyrrole compounds were analysed in detail (by Daniella Geissler, University of Applied Sciences, Aalen) using 337 nm laser. The pyrrole derivatives were found to be good matrices in general for small and large proteins in the positive ion mode and will be discussed in detail in the section 4.4.

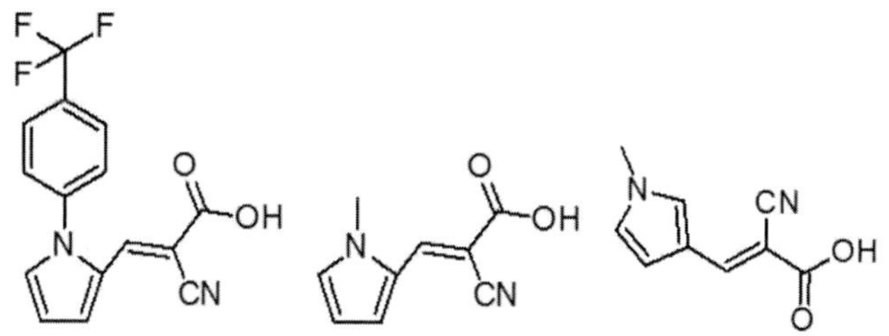

Figure 4.16: The structures of the matrices A) **65**, B) **83**, C) **87**

4.4 New matrices for Small and Large Proteins

4.4.1 Protein Calibration Standard-I using 337 nm (N_2) laser

The pyrrole derivatives were found to be excellent matrices especially for the peptides from BSA digest in positive ion mode. Some selected derivatives were tried as matrices for the detection of small and large proteins using them (by Daniella Geissler, University of Applied Sciences, Aalen).

The selected pyrrole derivatives (structures in Figure 4.16) were applied as matrices for the detection of small proteins from Protein calibration standard-I, (from Bruker GmbH, Germany). The standard mix consists of small proteins Insulin ($[M+H]^+$ = 5734.52 Da), Ubiquitin ($[M+H]^+$ = 8565.76 Da), Cytochrome C ($[M+H]^+$ = 12360.97 Da, ($[M+2H]^{2+}$ = 6181.05 Da), Myoglobin ($[M+H]^+$ = 16952.31 Da and ($[M+2H]^{2+}$ = 8476.66 Da)).

The spectra obtained by using pyrrole derivatives **65**, **83**, **87** are shown in Figure 4.17. The matrix solutions (5 mg/mL) were prepared in 1:1 ACN: H_2O, (v/v). The spectra (4000 shots summed) are compared to that of standard matrix sinapinic acid (20 mg/mL) in 1:1 ACN: H_2O, (v/v) for the protein standard mix solution (1 pmol/µL) in linear positive ion mode.

All of the selected 4 new matrices detect all the proteins from the mix. The signals obtained for the matrix **65** were intense but broad. The baseline was also distorted. The matrix **82** showed matrix cluster signals in the spectrum similar to the standard matrix sinapinic acid, although the signal intensities obtained with the new matrix are about three times better. The signals for matrix **87** were sharp and intense as compared to the sinapinic acid.

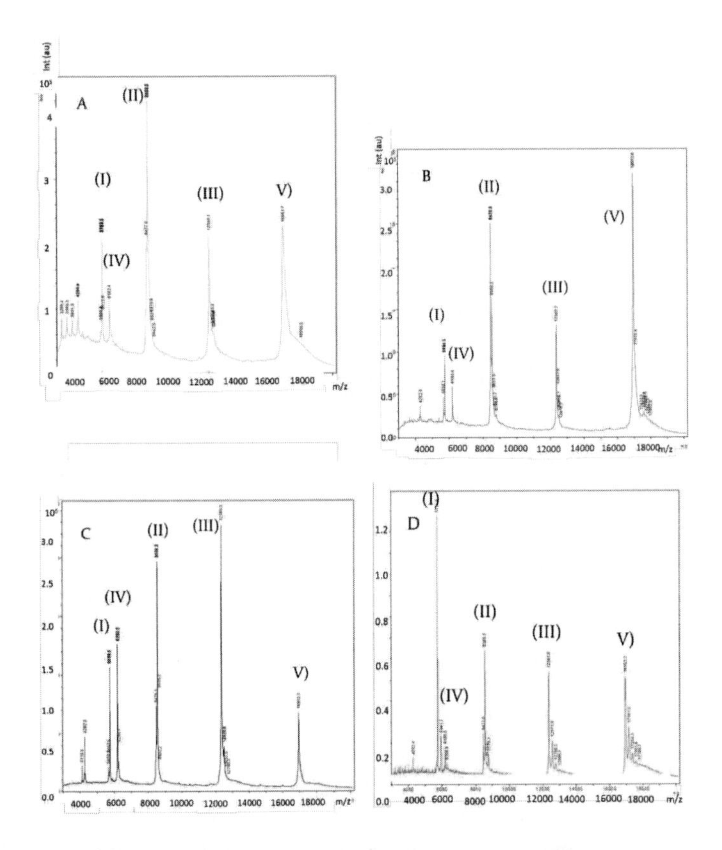

Figure 4.17: The comparison of mass spetra obtained for matrices A) **65**, B) **83**, C) **87** and D) Sinapinic acid for Protein calibration standard-I (1 pmol/µL) in positive ion mode using 337 nm laser. (I)Insulin ($[M+H]^+$ = 5734.52 Da), (II) Ubiquitin ($[M+H]^+$ = 8565.76 Da), (III) Cytochrom C ($[M+H]^+$ = 12360.97 Da, (IV) ($[M+2H]^{2+}$ = 6181.05 Da), (V) Myoglobin ($[M+H]^+$ = 16952.31 Da and (VI) ($[M+2H]^{2+}$ = 8476.66 Da). All matrix solutions were prepared in 1:1 ACN: H_2O, v/v

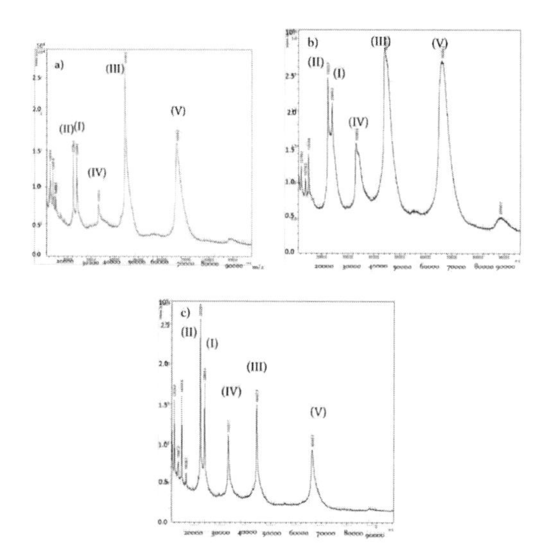

Figure 4.18: The comparison of mass spetra obtained for matrices a) Sinapinic acid, b) **65** and c) **82** for Protein calibration standard-II (4 pmol/µL) in positive ion mode using 337 nm laser. All matrix solutions were prepared in 1:1 ACN: H_2O, v/v

4.4.2 Protein Calibration Standard-II using 337 nm (N_2) Laser

Protein calibration standard-II (Bruker GmbH, Germany) consists of large proteins *viz.* Tripsinogen ($[M+H]^+$ = 23982 Da), Protein A ($[M+H]^+$ = 44613 Da, ($[M+2H]^{2+}$ = 22307 Da)) and BSA ($[M+H]^+$ = 66464 Da, ($[M+2H]^{2+}$ = 33232.5 Da)). The protein mix solution of 4 pmol/µL) was used for the evaluations with random walk. The spectra (4000 shots summed) were compared with the standard matrix sinapinic acid (20 mg/mL) and pyrrole derivatives (5 mg/mL) in 1:1 ACN: H_2O, (v/v).

The spectrum obtained by matrices 65 and 82 are shown in Figure 4.18. The matrix 82 showed sharp peaks in comparison to the matrix 65 and standard matrix sinapinic acid under the experimental conditions used.

4.4.3 Protein Calibration Standard-I using 355 nm (Nd:YAG) Laser

The protein calibration standard mix (Bruker GmbH, Germany) containing small proteins insulin, ubiquitin, cytochrom C and myoglobin was evaluated using some selected matrices (by Dr. Annabelle Fülöp, University of Applied sciences, Mannheim). The spectra were acquired with 500 fmol/µL) of the standard mix in both positive and negative ion mode. Total 1000 shots were summed (200 shots per step). The spectra obtained (see Figure 4.19 were compared with those from standard matrix sinapinic acid 20 mg/mL in 2: 1 ACN: H_2O, v/v + 0.1% TFA. All the new matrix solutions were prepared in Formic acid: H_2O: i-propanol (1:3:2) (v/v).

All the new matrices gave better resolution and higher signal intensities for the protein standard mix in both positive and negative ion modes.

4.4.4 Protein Calibration Standard-II using 355 nm (Nd:YAG) Laser

The protein calibration standard mix II (Bruker GmbH, Germany) made up of large proteins like tripsinogen, protein A and BSA was evaluated with some selected matrices (by Dr. Annabelle Fülöp, University of Applied sciences, Mannheim). The spectra were acquired for 1 pmol/µL) of the standard by collecting 1000 shots (200 shots per step) in positive ion mode. All the matrices solutions were prepared in Formic acid: H_2O: i-propanol (1:3:2) (v/v). All the new matrices performed comparable to that of sinapinic acid (see Figure 4.20)

4.5 p-PhCCAA a New Matrix for Lipid Classes in Negative Mode

Exciting results of matrix p-PhCCAA (**10**) on 355 nm laser led to detailed analysis of lipid extract in negative ion mode (synthesised by Miss Martina Porada, University of Applied Sciences, Aalen). The matrix **10** displayed matrix suppression effect (MSE) (explained in section 1.2.5.1) and less number of matrix signals and noise in the spectrum compared to the standard matrix 9-AA.

4.5.1 p-PhCCAA and Matrix Suppression Effect

Desorption and ionization of matrix molecules cause background peaks in the low-mass region (m/z < 1000) because of matrix fragmentation and cluster formation

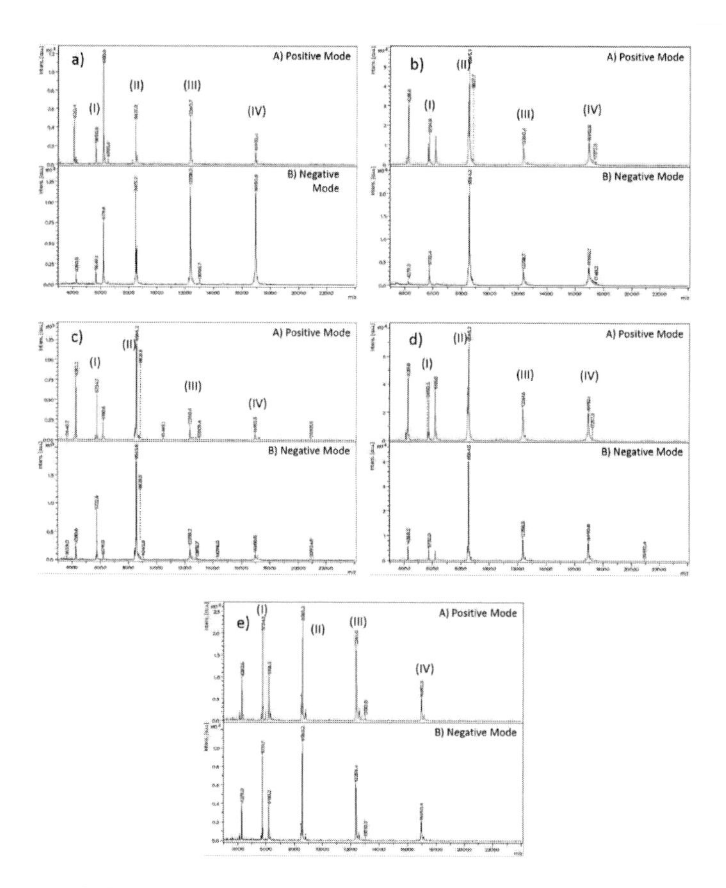

Figure 4.19: The comparison of mass spetra obtained for matrices a) **63**, b) **65** and c) **97, 98** and e) Sinapinic acid for Protein calibration standard-I (500 fmol/µL) in positive ion and negative ion modes using 355 nm laser.(I)Insulin ($[M+H]^+$ = 5734.52 Da), (II) Ubiquitin ($[M+H]^+$ = 8565.76 Da), (III) Cytochrom C ($[M+H]^+$ = 12360.97 Da, (IV) Myoglobin ($[M+H]^+$ = 16952.31 Da.

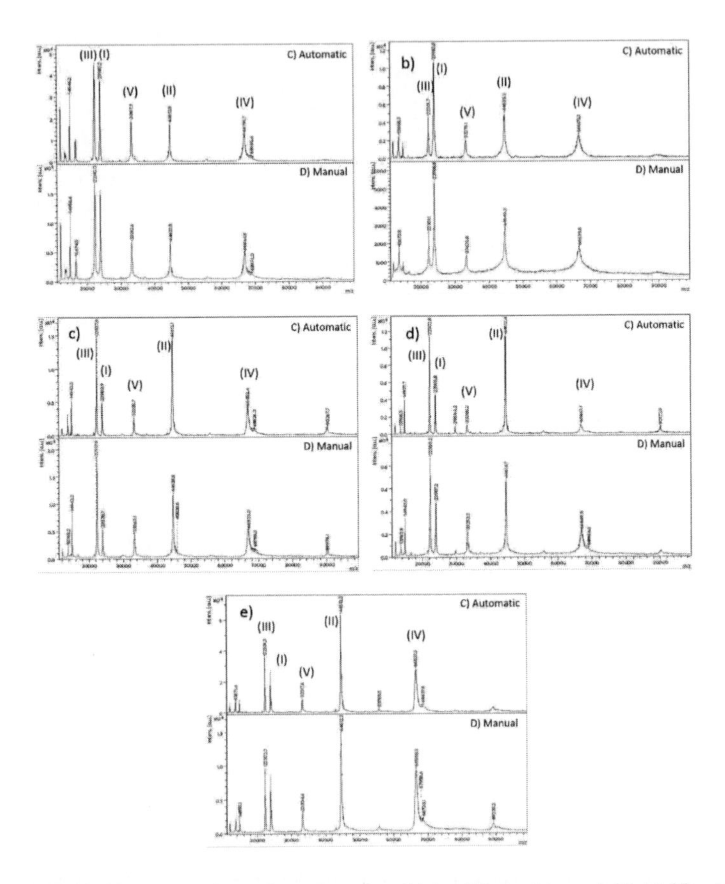

Figure 4.20: The comparison of mass spetra obtained for matrices a) **63**, b) **65** and c) **97** d) **98** and e) Sinapinic acid for Protein calibration standard-II (1 pmol/µL) in positive ion mode using 355 nm laser. (I) Tripsinogen ($[M+H]^+$ = 23982 Da), (II) Protein A ($[M+H]^+$ = 44613 Da, (III)($[M+2H]^{2+}$ = 22307 Da) and (IV) BSA ($[M+H]^+$ = 66464 Da, (V) ($[M+2H]^{2+}$ = 33232.5 Da)

during analyte detection. These background signals can be suppressed by sufficient analyte ions [124]. It has been noted that all the matrices do not show this MSE at the same ratio and is generally reduced if the crystallization is inhomogeneous [124]. The most easily adjustable factor is laser intensity, besides the matrix: analyte ratio. As more matrix ions are generated by stronger laser pulse, that require more analytes to quench them, the MSE is maximal near the MALDI laser intensity threshold [125] Nevertheless, it is practically impossible to prepare samples with an appropriate matrix:analyte ratio, since the analyte concentrations in biological samples are often unknown. To overcome this limitation, if the matrix of choice already promotes the MSE at low amounts of sample, it is desirable [125].

The background matrix signal formation and their suppression by matrices p-PhCCAA (**10**) and 9-AA (**60**) was examined (by Dr. Annabelle Fülöp, University of Applied Sciences, Mannheim, [97]) near laser intensity thresholds with and without 20 ng of BTLE. In the absence of lipid extract, characteristic matrix cluster formation was observed in the case of 9-AA, whereas the matrix spectra of p-PhCCAA (**10**) showed fewer background signals (see Figure 4.21). The majority of the matrix signals for **10** were suppressed in the presence of lipid extract. This MSE was not observed in the case of 9-AA matrix with 20 ng of lipid extract. The MSE_{lip} score with 20 ng extract was 4 times improved with **10** when compared to 9-AA.

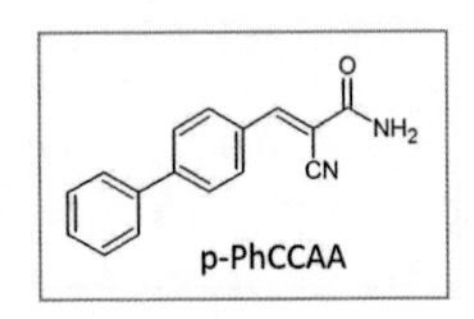

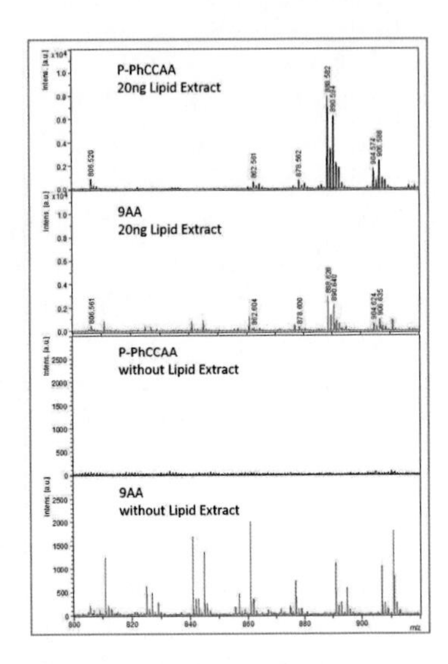

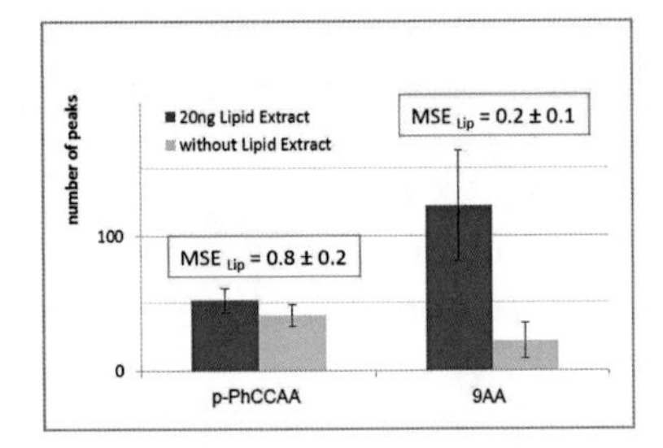

Figure 4.21: The comparison of spectra with and without brain total lipid extract (BTLE) using matrices p-PhCCAA (5mg/mL in 90% acetone and 9-AA (5mg/mL in 3:2 IPA: ACN, v/v) in negative ion mode of MALDI-MS using 355 nm Nd:YAG laser (adapted from [125]).

4.5.2 p-PhCCAA and MALDI-Imaging

It was reasoned therefore that the small number of background peaks in the m/z range 400−2000 Da may imply that p-PhCCAA be useful in the combined detection of sulfatides and other lipids in IMS experiments [125]. The IMS data set using rat brain cryosections was acquired (by Miss Annabelle Fülöp, University of Applied Sciences, Mannheim). In which one piece of tissue was coated with 9-AA and the other with p-PhCCAA. The Figure 4.22 shows imaging data when the 9-AA-coated region was measured first. The reported selectivity of 9-AA for sulfatides [SM4s(42:2) and SM4s(h42:2)] was confirmed [95; 126]. The new matrix p-PhCCAA (**10**) also enabled imaging and localizing other lipids such as phosphatidyl inositols [e.g., PI(36:4) and PI(38:4)] and even the phosphatidyl ethanolamines [PE(38:4) and PE(p38:4)] or the phosphatidyl glycerol species [PG(34:1) and PG(40:6)].

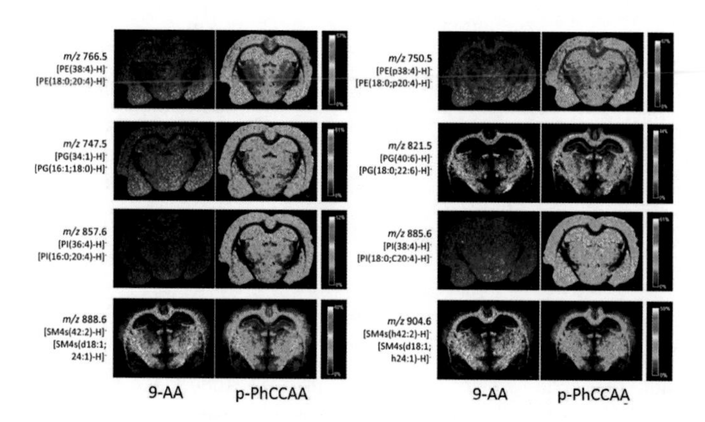

Figure 4.22: p-PhCCAA matrix enables MALDI-IMS of various lipid species. Imaging was performed on two adjacent cryosections of SpragueDawley rat brain on the same ITO slide. Data was acquired in negative ion-reflector mode at a spatial resolution of 100 µm with 200 laser shots per position. Prior to measurement, the same amount of 9-AA [5 mg/mL in acetonitrile/water (80:20, v/v)] (left side) or p-PhCCAA [5 mg/mL in acetone/water (90:10, v/v)] (right side) was deposited onto one of the tissue sections with a SunCollect MALDI spotter. Imaging was carried out while the laser energy was kept constant and slightly above the individual threshold level for one region. The 9-AA region was measured first. Mass filters were chosen with a width of 0.2 Da. (Adapted from [125]

4.6 Structure-Performance Relationship of Phenylcinnamic Acid Derivatives

For MALDI-MS matrix evaluation p-PhCCAA matrix (10) and 9-AA were used as reference matrix compounds and BTLE was chosen as a reference analyte mixture [125]. This lipid extract is well examined and it consists of a mixture of different lipids classes like gangliosides, phosphatidylcholines (PCs), phosphatidylethanolamines (PEs), sulfatides (STs), phosphatidylinositols (PIs), phosphatidylserines (PSs), phosphatidic acids (PAs) and phosphatidylglycerols (PGs) [127]. Sulfatides were chosen as a representative lipid class for the evalution of phenyl-α-cyanocinnamic acid derivatives as potential matrices in the negative ion mode. Out of this class 14 individual sulfatides were assigned based on literature data [127; 128]. This class is especially suited for

MALDI-MS matrix evaluation in negative ion mode due to the large dynamic range of the observable S/N ratios and the multiplicity of structurally related members. There are also some members of this class of lipids, which are relatively abundant in this lipids mixture, allowing also the evaluation of comparable poor performing matrices.

4.6.1 Performance Dependence on the Molar Extinction Coefficient

A measure which is not biased for the absolute S/N ratios for individual sulfatides must be defined so as to compare the relative performance of each matrix for sulfatide detection, which reflect their natural abundance and their ionization efficiency in the BTLE. This measure of the performance of an individual matrix was viewed in terms of the relative mean S/N ratio for each matrix (see Figure 4.23. The relative mean S/N ratio was calculated by dividing the sum of the relative S/N values through the number 14 of assigned sulfatides as shown in the following equation.

$$\text{Relative mean S/N ratio} = \frac{1}{14} \sum_{i=1}^{14} \frac{(\text{S/N matrix sulfatide species})_i}{(\text{S/N 9-AA sulfatide species})_i} \qquad (4.1)$$

Compound **60** is 9-AA with the reference performance of 1 recorded using a 337 nm laser and a relative performance of 0.5 for 9-AA at 355 nm in comparison to 337 nm. The difference in the relative mean S/N ratios obtained at two different laser wavelengths were comparatively higher for the compounds **10**, **12**, **17**, **22** and **40** than for the remaining 54 matrices (see Figure 4.23).

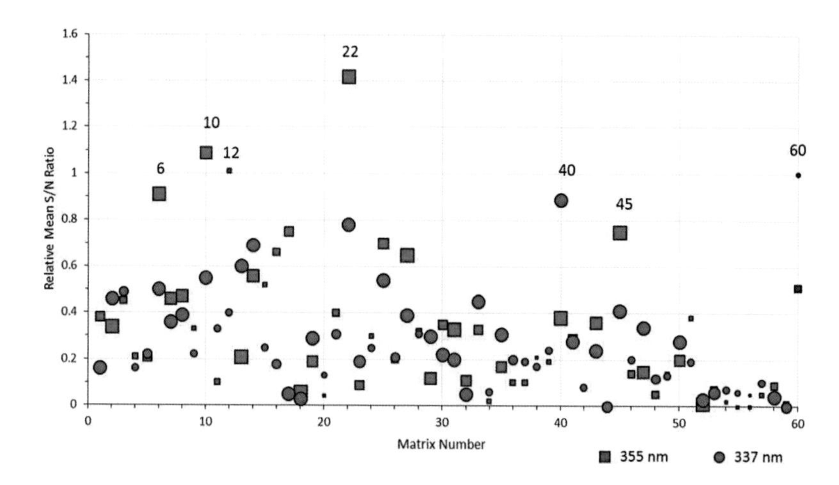

Figure 4.23: The performance dependence on the molar extinction cofficients of the matrices. The size of the spots increase with increase in the molar extinction coefficients. (green square: 355 nm, blue circle: 337 nm, black square: 9-AA at 355 nm and black circle 337 nm)

Energy absorption at the laser wavelength is essential for matrix performance. Figure 4.23 shows a positive correlation between the molar extinction coefficient ε_{355} or ε_{337} in solution and the MALDI-MS performance of most of the matrices, an expected result assuming that molar UV-Vis absorbance properties in solid phase and in solution phase are coarsely correlated [129; 130]. For example, matrices **10** and **22** with an outstanding MALDI-MS performance display high molecular extinction coefficients in solution phase. To examine the differences between solid and liquid phase UV-Vis absorbance solid phase UV-Vis spectra were acquired for some selected matrices (**10**, **11**, **12**, **22**, **23**, **24**, **19** and **25**, Table 4.1). The only approximate correlation between solution phase extinction and matrix performance (4.1) can be in partly explained by the observed peak broadening and bathochromic shifts for λ_{max} values in solid state (Figure 4.24) leading to higher absorbance at 355 nm in solid state compared to solution phase.

4.6 Structure-Performance Relationship of Phenylcinnamic Acid Derivatives

Matrix	Structure	λmax solid	ελmax sol	RMSN-355nm	RMSN-SD-355 nm (%)	ε337nm sol	RMSN-337nm	RMSN-SD-337nm (%)
10	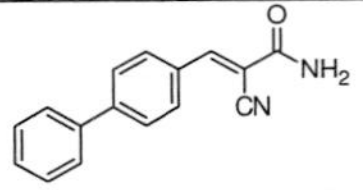 O, NH_2, CN	354 ± 0.6	18144	1.1	6	27958	0.6	17
11	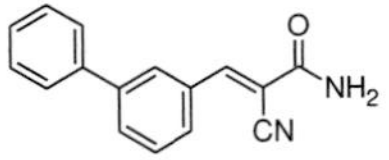 O, NH_2, CN	329 ± 2.6	967	0.1	31	3751	0.3	11
12	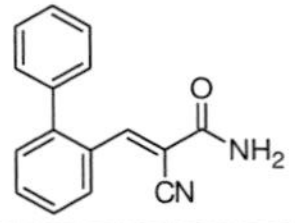 O, NH_2, CN	329 ± 1.0	257	1	11	3196	0.4	7
22	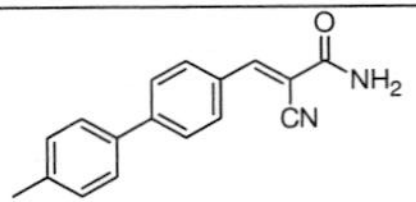 O, NH_2, CN	346 ± 0.0	23932	1.4	9	28937	0.8	6
23	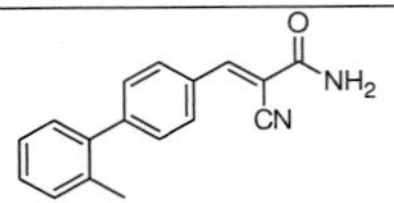 O, NH_2, CN	346 ± 1.0	7795	0.1	18	19191	0.2	17
24	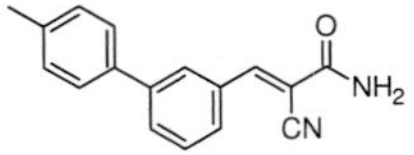 O, NH_2, CN	318 ± 4.0	450	0.3	14	2560	0.3	11
19	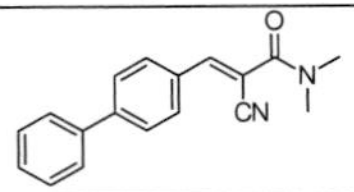 O, N, CN	351 ± 0.6	11206	0.2	19	24398	0.3	6

Matrix	Structure	λmax solid	ελmax sol	RMSN-355nm	RMSN-SD-355 nm (%)	ε337nm sol	RMSN-337nm	RMSN-SD-337nm (%)
43	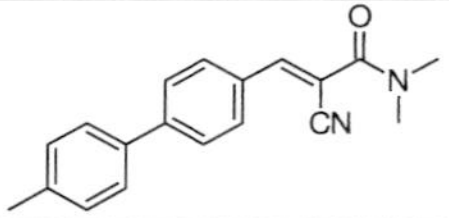 O, N, CN	-	17949	0.4	7	28197	0.2	8
25	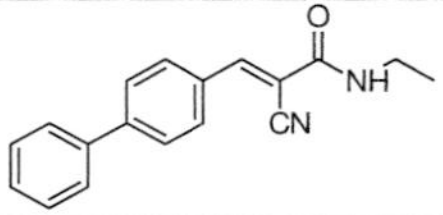 O, NH, CN	353 ± 1.5	13450	0.7	7	24972	0.5	6
45	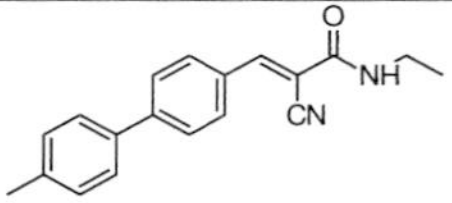 O, NH, CN	-	23879	0.7	11	28512	0.6	7
60	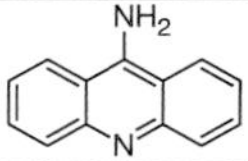 NH_2, N	281 ± 1.0	1641	0.5	6	413	1	6

Table 4.1 Selected geometrical isomers with their structures, molar extinction coefficients in [L/molcm], relative mean S/N ratios and mean standard deviations, calculated by averaging the relative standard deviations for 14 sulfatides. (sol: solution, RMSN: relative mean S/N ratio, SD: standard deviation).

4.6.2 Performance Dependence on the Substitution Position

Other than the UV-Vis absorption properties of the matrices and their performances a relationship between relative mean S/N ratios and phenyl substitution pattern of the cinnamic acid core is suggested by the analysis of the data (Figure 4.25). A comparison of the solution phase extinction coefficients of constitutional isomeric matrices **10-12** and **22-24** and their relative mean S/N ratios along with their mean relative standard deviations in percentage is shown in the table 4.1. A comparison of the performances of three isomeric matrices with Scaffold a (**10**), Scaffold b (**11**) and Scaffold c (**12**) showed that the para-isomer (**10**), (Scaffold a) showed the best relative mean S/N ratios for sulfatide detection in the negative MS ion mode. This finding was as expected because of the higher ε value (ε_{355} = 18144 [L/(mol cm)] and ε_{337} = 27958 [L/(mol cm)]) of **10** compared to matrices **11** and **12**. Surprisingly, in contrast, the matrix **12**, an ortho phenylcinnamic acid derivative displayed a better relative mean S/N ratio in comparison to matrix **11**. This reult was not consistent with the absoption coefficient values in solution (ε_{355} = 967 (**11**) and 257 (**12**) [L/(mol cm)] and ε_{337} = 3751 (**11**) and 3196 (**12**) [L/(mol cm)]) and solid state λ_{max} (Figure 4.24).

Similarly, in another example matrix **22**, a p-(4-methylphenyl)cinnamic acid derivative (Scaffold a), displayed significant highest relative mean S/N ratios. In comparison, the p-(2-methylphenyl) substituted cinnamic acid derivative **23** (Scaffold a) displays even lower relative mean S/N ratios than the m-(4-methylphenyl)cinnamic acid derivative **24** (Scaffold b). This suggest a role of the substitution pattern on the aromatic ring and the steric hindrance caused by the substition on the conformation of the compound. In this example, the methyl subsituent causes steric hindrance in **23** compared to **22** which leads to the poorer performance, indicating a possible role of compound conformation for MALDI matrix performance.

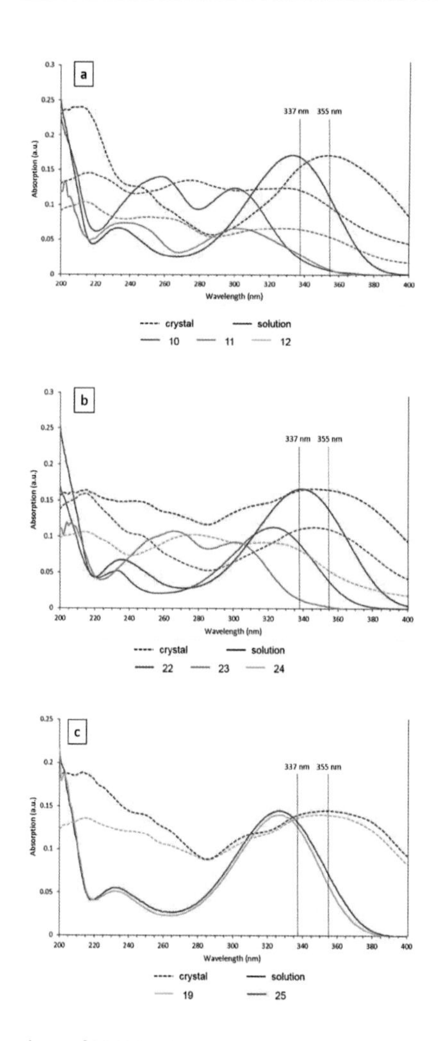

Figure 4.24: Comparison of UV-Vis absorption of matrices a) **10**, **11** and **12** in solution phase (Methanol : Water, 4:1, v:v) and solid state. ($A_{solution}$ for matrix **10**: 2.5A_{solid}; matrix **11** : 0.3A_{solid}; matrix **12**: 0.16A_{solid}) b) **22**, **23** and **24** in solution phase (Methanol : Water, 4:1, v:v) and solid state. ($A_{solution}$ for matrix **22** : 0.69A_{solid}; matrix **23** : 0.37A_{solid}; matrix **24** : 0.16A_{solid}) c) **19** and **25** in solution phase (Methanol : Water, 4:1, v:v) and solid state. ($A_{solution}$ for matrix **19** : 0.69A_{solid}; matrix **25** : 1.08A_{solid})

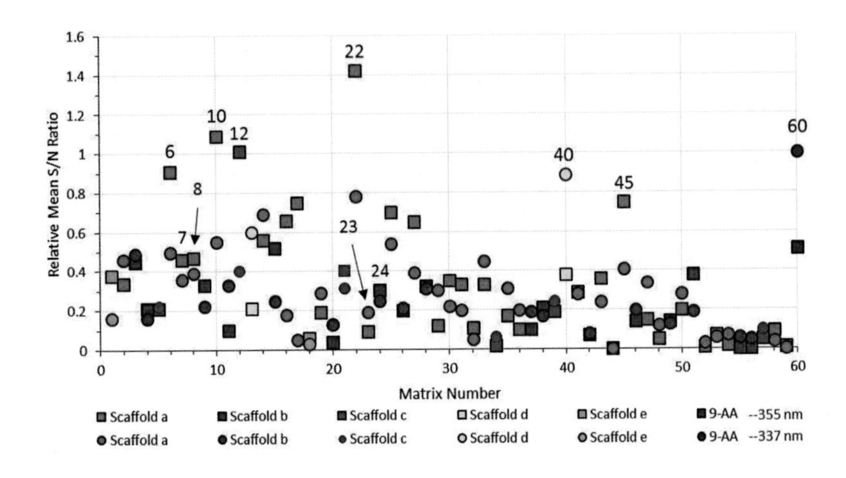

Figure 4.25: The performance dependence for different geometries of matrices. The color of the spots are different for different geometry (Scaffold a = dark green; Scaffold b = red; Scaffold c = blue; Scaffold d = pale green; Scaffold e = orange; 9AA = black) and the shapes are different for the laser wavelengths (Square = 355 nm and Circle = 337 nm)

4.6.3 Performance Dependence on the Functional Group at Carboxy Carbon

To analyse the the dependence of the performance of the matrices on the funtional groups at the carboxy carbon, the data was analysed for acids, primary amides, mono- and di-alkylated amides and the acetyl urea derivatives. It has already been established that the amide derivatives of phenyl-α-cyanocinnamic acids are in general better MALDI-MS matrices than the corresponding acids using the negative ionisation mode [97] (Figure 4.26). Consistent with this result, out of 24 amide-acid pairs evaluated, amides performed better than their corresponding acids in 21 acid-amide pairs. It was observed that the performance varies significantly when the substituents are introduced at the primary amide indicating a possible role of the amide hydrogens for H- bonding or acid/base chemistry. Data analysis confirmed the importance of both the amide nitrogen-hydrogens. An order of performance was established as follows: primary amide > ethylamide > acetylurea > acid > dimethylamide in a group of acid and amide derivatives **10**, **25**, **2**, **33** and **19** at 337 nm laser wavelength. Replacement of the hydrogens from amide nitrogen reduces the possibility to form hydrogen bonds that might be the possible reason for the performance modulation. It is worth mentioning here that, a comparison of solution phase and solid state UV-Vis measurements of **19** and **25** showed that UV-Vis properties of these compounds are not greatly altered by introduction of the methyl groups at the amide nitrogen, but the performance alters dramatically.

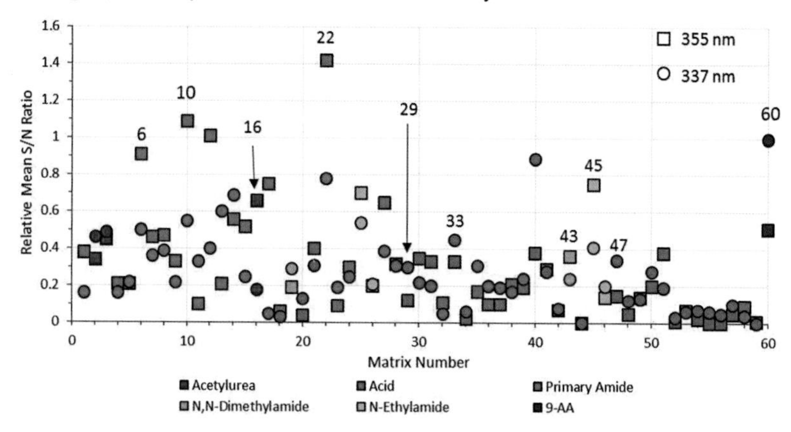

Figure 4.26: MALDI-MS performance dependence on functional group at carboxy position: the color of spots are different for different functional groups (acid = blue; amide = dark green; *N*-ethylamide = pale green; *N,N*-dimethylamide = orange; acetylurea = red and 9-AA = black). The shapes are different for the laser wavelengths (Square = 355 nm and Circle = 337 nm)

4.6.4 Selectivity and Performance Dependence on the Hydrophobicity of the Solvent System

The Figure 4.27 compares the S/N ratios (355 nm) for a representative sulfatide SM4s(42:1)] against matrix logP values in solvent systems with different hydrophobicities. The results presented are for selected matrices **2**, **6**, **10**, **12**, **16**, **19**, **33**, **43**, **46** and **47** only.

To analyse whether any of the 59 matrices exhibit selectivity for any particular sulfatide, a measure is needed that could relate individual performances of the matrices for each sulfatide separately. To address this issue, the best performing matrix

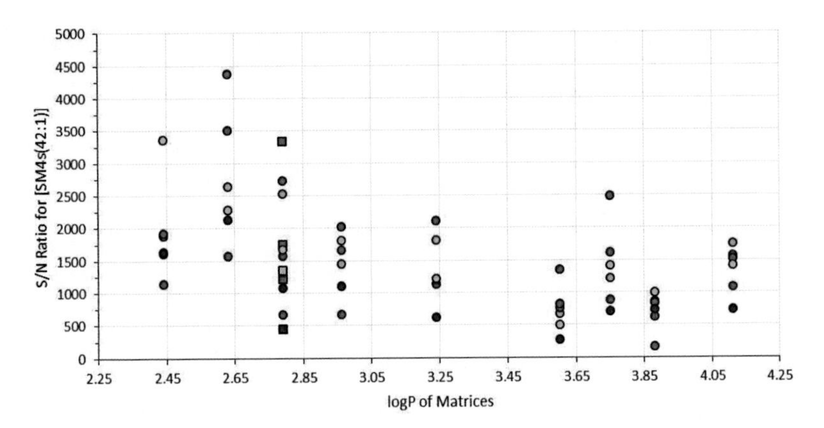

Figure 4.27: The dependence of S/N ratios of a representative sulfatide [SM4s(42:1)] on logP of matrix in different solvent systems. The selected matrices numbers were **2**, **6**, **10**, **12**, **16**, **19**, **33**, **43**, **46** and **47**. Isomer matrices **10** and **12** (shown by squares) have same logP values

22 was used as the reference compound. The S/N ratios obtained for each matrix for each sulfatide relative to S/N ratio obtained using matrix **22** was calculated by dividing S/N ratio-matrix by S/N ratio-matrix 22 separately for 355 and 337 nm lasers. In this way, the relative S/N ratio for matrix **22** was always 1 for all the sulfatides and any matrix showing > 1 relative S/N ratio for any specific sulfatide would be selective for that particular sulfatide. Figure 4.28a/b, show the results for 355 and 337 nm respectively. It shows that there is no special selectivity exhibited by any matrix.

4.7 X-ray Crystal Structure Analysis of Selected MALDI-Matrices

The (*E*) geometry of the double bond is confirmed by the X-ray crystal structure packings of the matrices **10**, **22**, **11**, **19**, **26** and **45** visualised in the Figure 4.29. The crystal packing patterns of **10**, **11** and **22** show hydrophobic regions and hydrogen bond interactions of the amide functionalities. Crystal structures of primary amides **10** and **22** show the presence of H-bonds between R-C≡N|· · ·H-NR_2 and R_2N-H· · ·|O=CR_2 functionalities, such that each molecule of **10** and **22** is connected to two adjacent matrix molecules by hydrogen bonds. In case of primary amide **11**, an ortho derivative, two independent molecules form dimeric structures by hydrogen bonds between amide residues (R_2N-H· · ·|O=CR_2), while the cyano group does not participate in hydrogen bonding, as in the matrices **10** and **22**. Figure 4.29 shows the differences in the H-bonding pattern for the *N*-ethylamides i.e. if one of the amide hydrogens is replaced by an ethyl group. In the para-phenyl substituted (Scaffold a) compound **45**, hydrogen bonds are between amide residues (R_2N-H· · ·|O=CR_2) of two different molecules similar to the matrix **10** and **22**. In meta-phenyl substituted (Scaffold b) *N*-ethyl derivative (**26**), a dimer is formed via hydrogen bonding between cyano and amide NH groups (R-C≡N|· · ·H-NR_2). The crystal structure of *N*,*N*-dimethylamide derivative (**19**) where both the amide hydrogens are replaced by the methyl groups, shows absence of hydrogen bonding.

The sequential replacement of the amide hydrogens from p-PhCCAA (**10**) towards the *N*-ethyl- (**25**) and *N*,*N*,-dimethyl (**19**) naturally reduces the number of possible H-bonding interactions. Figure 4.26 displays that when these number of H-bonding interactions are reduced from primary amide (**10**), *N*-ethyl- (**25**) and *N*,*N*-dimethyl (**19**), their performance also diminishes. This points out the possibility of the role of H-bonding interactions on the performance of matrices.

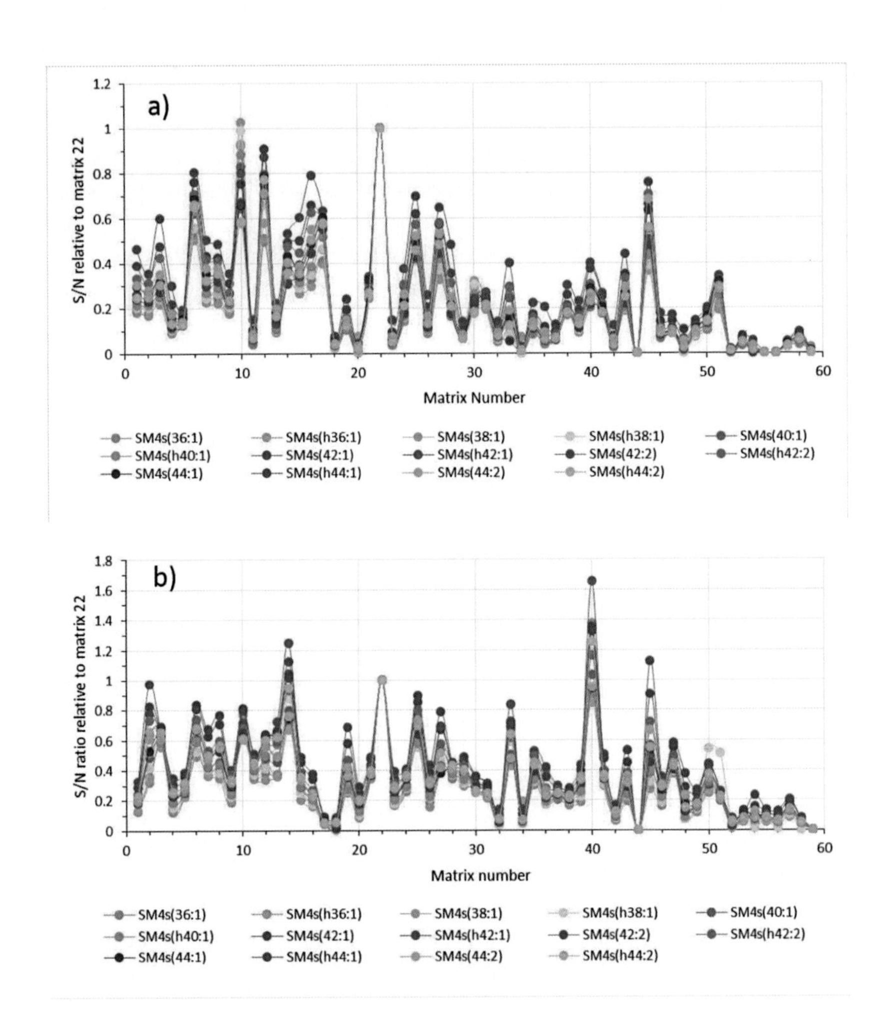

Figure 4.28: Selectivity of matrices for individual sulfatides using a) 355 nm laser and b) 337 nm laser. The S/N ratios for individual sulfatides were calculated relative to S/N ratios obtained by matrix **22**

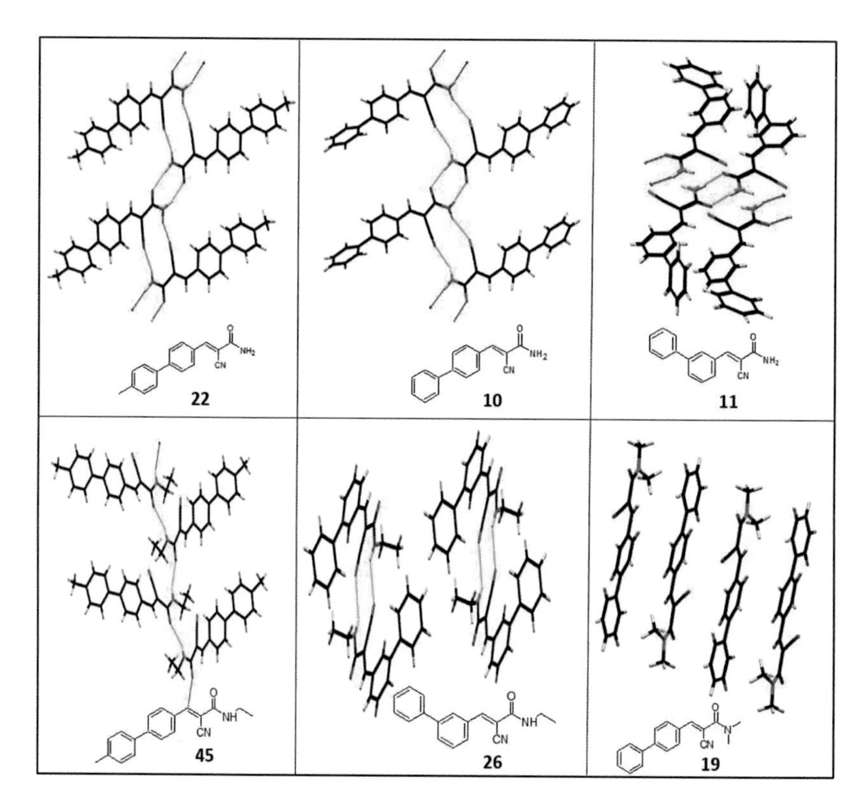

Figure 4.29: Crystal structures of matrices **22**, **10**, **11**, **45**, **26** and **19** showing differences in H-bonding pattern

A similar trend was noted for another set of matrices primary amide **22**, *N*-ethyl-**45** and *N*,*N*,-dimethyl **43**. The compounds **43** and **45** are constitutional isomers with closely related physical properties like hydrophobicity and UV absorbance (Table 1). The *N*,*N*-dimethyl derivative (**43**) has on the contrary a 30°C lower melting point in comparison to **45** indicative of a lower crystal lattice energy due to the lack of stabilizing hydrogen bond interactions.

4.8 Principal Component Analysis

The detailed introduction to the principal component analysis (PCA) and interpretation of PCA are given in the section 2.3.1. The principal components are the dimensionless linear combinations of the original data set that are used to reduce complexity. The highest variance is shown by the first component (denoted as PC1). The following components have gradual decrease in their variance and are termed as second component (PC2), third component (PC3) and so on which are orthogonal to the PC1. Scree plot explains how many components have significance on the PCA. PCA can be useful tool to study the roles of large number of calculated and experimentally determined parameters on the performance of the matrices.

The orientation of the variable vectors in the loading plot indicates the possible correlations among the parameters. The length of the parameter vector explains the variance specified by the principal components. In other words, the length is a measure of the influence or impact of the parameter on the PCA. If the vectors for different parameters point out in the same direction then the parameters have positive correlation between them. If they point out in opposite directions, then they are negatively correlated and if orthogonal to each other then they indicate no correlation.

Besides structural features of MALDI-matrices general physical properties can be of importance for matrix performance. A PCA was carried out to uncover performance-relevant matrix properties out of a large set of 17 structural and physical parameters as the variables. The numbers are assigned to each parameter for the sake of clarity and to simplify the figure as follows: 1 = logP 2 = pKa 3 = H-donor 4 = H-acceptor 5 = molecular weight 6 = ε_{355} 7 = ε_{337} 8 = melting point 9 = relative mean S/N-355nm 10 = relative mean S/N-337nm 11 = laser power-337nm 12 = number of exchangeable protons 13 = lone pair of electrons on head group 14 = substitution position 15 = biphenyl connectivity 16 = laser power-355 nm and 17 = number of atoms on the amide head group. The values assigned for biphenyl connectivity were 2 = Scaffold a, 1 = Scaffold b and 0 = Scaffold c; and for substitution position 6 = no substituent, 4 = para-position, 3 = meta-position, and 0 = ortho-position. Thus, bigger the value of the parameter, smaller is the steric hindrance within the molecule caused by aromatic rings and different substituents on the rings. PCA delivers hints or trends for possible underlying correlations in large data sets and must be validated by direct analysis of the original data.

The scree plot of the PCA is shown in Figure 4.30, in which first five components were of significant importance (eigenvalue ≥1). Figures 4.31 and 4.32 present the loading and score plots of the PCA carried out for all 59 biphenyl derivatives for first three principal components in 3-dimensional representation. The analysis of the PC4 and PC5 were also carried out with different combinations of the components, but are not presented here as they did not reveal any additional information about the correlations. In the loading plot, a two-dimensional correlations of the variable vectors of any two components are also projected for the sake of clarity (in red colour). The variance of 1st component was 27.5% , that of 2nd component was 21.1% and for the third component variance was 14.4%, giving total variance of 63%.

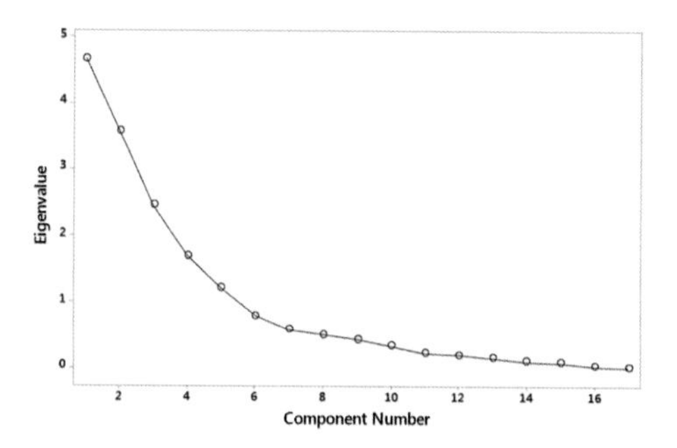

Figure 4.30: Scree plot of the principal component analysis

The loading plot between the first three components of the PCA delivers possible relationships among matrix properties. Figure 4.31 shows that each of the parameters logP values (parameter 1), melting points (parameter 8) and molecular weights (parameter 5) of the matrices are located in opposite direction to the relative mean S/N ratios for the sulfatides for two lasers (parameters 9 and 10), implying a negative correlation. Molar extinction coefficients at 337 and 355 nm (parameters 6 and 7 are pointing in the same direction as that of relative mean S/N ratios (parameters 9 and 10) but the angle between them is large. This observation suggests that it is positively correlated parameter for the performance, but it is not the sole parameter. It is possible that altough molar extinction coefficients are high, the compound might not be a good matrix.

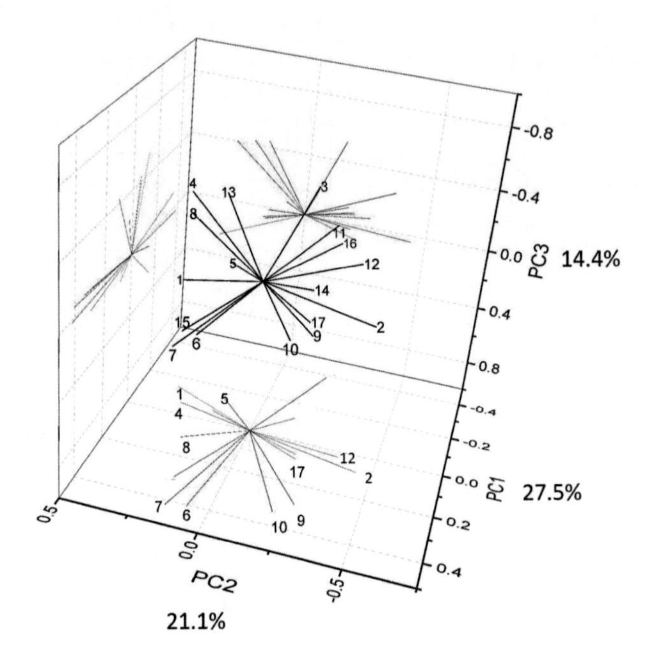

Figure 4.31: Principal Component analysis: Loading Plot: 1 = logP 2 = pKa 3 = H-donor 4 = H-acceptor 5 = molecular weight 6 = ε_{355} 7 = ε $_{337}$ 8 = melting point 9 = relative mean S/N-355nm 10 = relative mean S/N-337 nm 11 = Laser power-337 nm 12 = number of exchangeable protons 13 = lone pair of electrons on head group 14 = substitution position 15 = biphenyl connectivity 16 = laser power-355nm 17 = Number of atoms on the head group

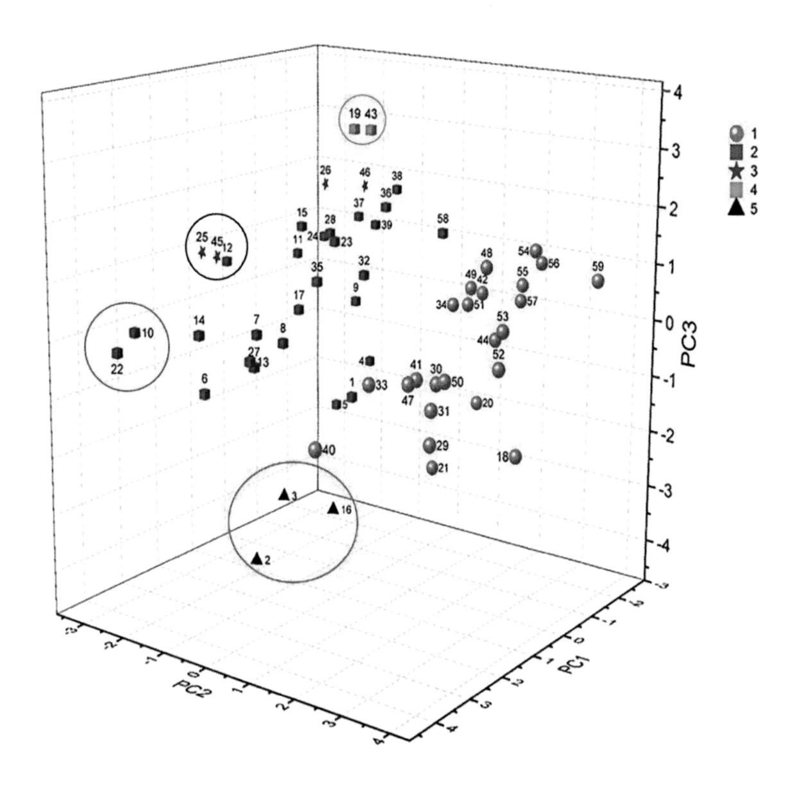

Figure 4.32: Principal Component analysis: Score Plot: Different functional groups are shown by different colours and shapes (1 = acid 2 = amide 3 = *N*-ethylamide 4 = *N*,*N*-dimethylamide 5 = acetylurea)

Additionally, pKa (parameter 2) and number of exchangeable protons (parameter 12) are located in the same direction like the relative mean S/N ratios for the sulfatides (parameters 9 and 10) implying a positive correlation between these values. The H-bond donor numbers and the optimal laser powers used with two different lasers are orthogonal, a potential correlation is not verifiable. The H-bond donor numbers and the melting points of matrices (because melting may change the phases in the crystals) presumably do not contribute significantly to the performance as they are orthogonal.

The high impact of the S/N ratios at the two different wavelengths on the first principle component makes a comparison of loading and score plot necessary to reassure the results. The score plot (Figure 4.32 uncovers relationships along individual matrices. The observed variable relationships were validated by similar regions for amides and amide derivatives in the score plot compared to the direction of S/N ratios in the loading plot. Figure 4.26 shows that these matrices have very good S/N ratios at both laser wavelengths. Score plot shows structurally related compounds grouped together. *N*-ethylamides (**25** and **45**, black circle), acetylurea derivatives (**2**, **3** and **16**, red circle), *N*,*N*-dimethylamides (**19** and **43**, orange circle) group nicely. The best performing matrices **10** and **22** are group together (blue circle) and are towards extreme left of the score plot, followed by the next best performers (**6**, **14**, **12**, **25** and **45**). Similarly, many bad matrices are grouped together on the right side of the score plot, most of them are acids, confirming the observation that amides are better as matrices than acids.

Due to the comparatively small data set, a hierarchical cluster analysis performed with the complete linkage method for verification purposes. The corresponding dendrogram (4.33) shows a comparable grouping of matrices supporting these results, as well. Cluster analysis displayed similar groupings as shown in the score plot of PCA.

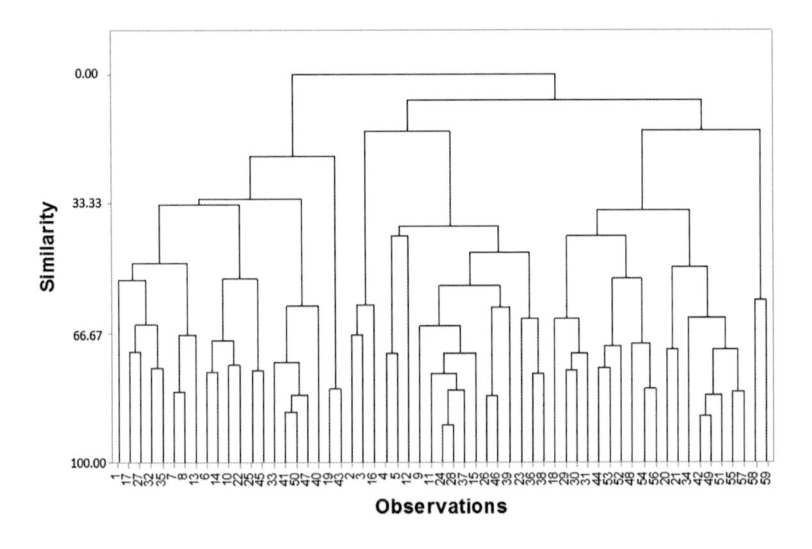

Figure 4.33: The hierarchical cluster analysis with the complete linkage method

The trends obtained from the PCA should be confirmed from the real data from the MS-measurements. The PCA suggested negative correlation between logP values and molecular weight respectively with the relative mean S/N ratios. Figure 4.34 shows the dependence of the S/N ratio for a representative sulfatide [SM4s(42:1)] on the logP values and molecular weights separately using 355 nm laser.

It is evident that as the logP values decrease, S/N ratio increase. For smaller logP values (less than 2.5) the S/N ratios were small again. The negative correlation of the molecular weights with the S/N ratios was confirmed (Figure 4.34b). S/N ratios get better and better as the molecular weights get smaller and smaller. But this trend did not work for all the compounds as there are many other parameters that influence the performance other than molecular weights.

4.9 Summary of the Chapter

- Discussed the results obtained by using different types of matrices, i.e. HCCA derivatives, diene derivatives, pyrrole derivatives and phenylcinnamic acid derivatives; for the detection of peptides either from BSA digest or

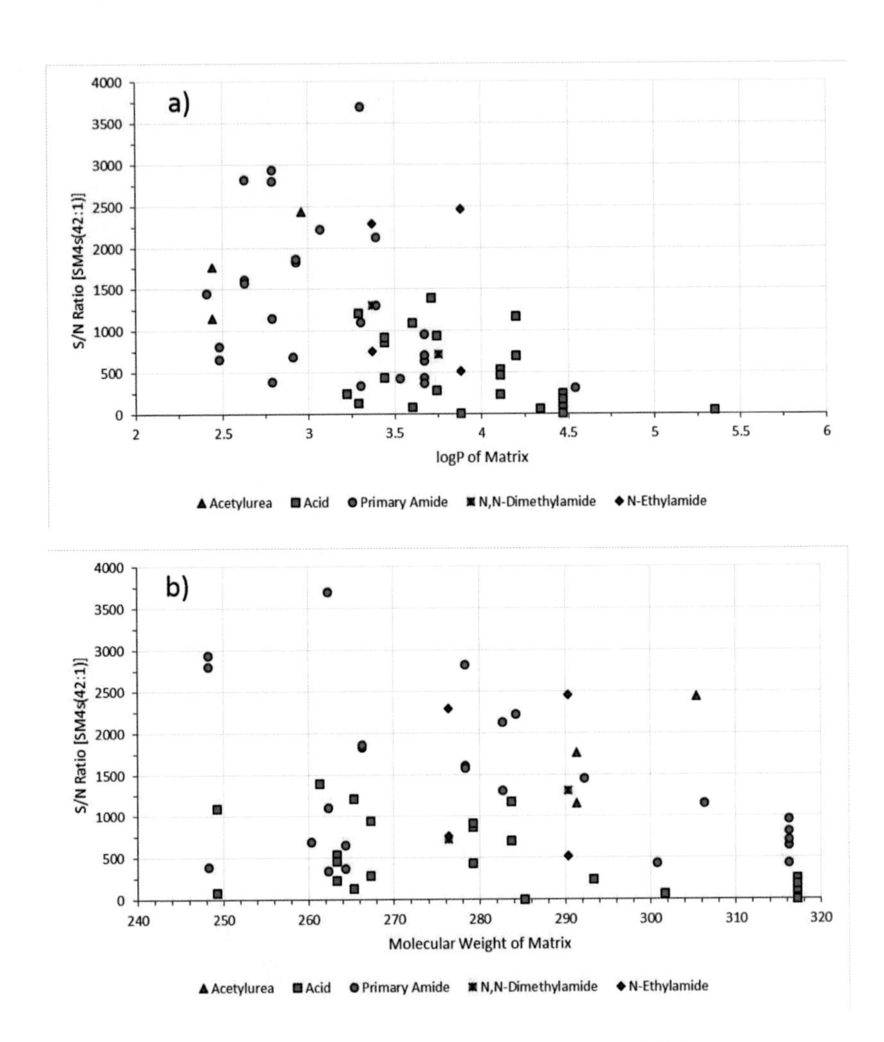

Figure 4.34: The dependence of S/N ratios of a representative sulfatide [SM4s(42:1)] on a) logP values of matrices and b) molecular weights of the matrices

that of the lipids from BTLE, protein calibration standard mixtures containing small and large proteins and compared them with the standard matrices

- Discussed the results obtained by using phenylcinnamic acid derivatives as MALDI matrices for lipids, especially sulfatides, in negative ionisation mode of mass spectrometry.
- Compared the results obtained by using phenylcinnamic acid derivatives for two different lasers with different laser wavelengths (337 and 355 nm)
- Several structural features of the matrix were revealed to be vital for its performance. This is a significant step towards a rule-based identification process of MALDI-matrix.

5

Conclusion

The goal of this chapter is to put all the results obtained during this doctoral study together, in brief.

The development of matrix is an issue of trial and error. Some parameters like molar extinction coefficients, pKa and hydrophobicity are studied to some extent. But still the rule based development of the MALDI-MS matrix has not been the reality till date.

To address this issue a compound library of about 200 compounds (cinnamic acid derivatives) was designed and synthesised. These potential matrices were used for the detection of peptides from BSA digest and lipids from BTLE in positive and negative ion mode respectively.

A novel matrix from this class was developed rationally for 355 nm Nd:YAG laser for lipid analysis in negative ion mode of mass spectrometry. The replacement of hydroxyl group of HCCA by proton affine substituent like chlorine leads to hypsochromic shift. This makes Cl-CCA unsuitable for the Nd:YAG laser. To approach to avoid this hypsochromic shift led to the replacement of the hydroxy group of HCCA by a phenyl ring.

The new matrix synthesised with this rationale (by Miss Martina Porada, University of Applied Sciences, Aalen) p-PhCCAA (**10**) exhibited extraordinary matrix properties for the lipids for 355 nm laser in negative ion mode. The matrix **10** displayed features of MALDI matrix with improved sensitivity and reproducibility. It also displayed matrix suppression effect and smaller number of background peaks in the mass range of lipids compared to standard matrix 9-AA. Its homogeneous crystallisation allowed its use for the imaging purposes. As compared to the selectivity of 9-AA for sulfatides, the new matrix (**10**) resulted for imaging on two adjacent cryosections from the Sprague–Dawley rat brain in better signal intensities for different lipid classes like PEs, PGs, PIs and SMs 4.22 [97].

These result initiated our interest to synthesise several phenylcinnamic acid derivatives. A compound library of 59 phenylcinnamic acid derivatives was synthesised and these potential matrices were evaluated for the detection of lipids from BTLE in negative ion mode. The evaluations were carried out on two most commonly used lasers *viz*. 337 (N_2) and 355 nm (Nd:YAG) lasers. The comparison of the measurements for 337 and 355 nm lasers for different structural and physical properties of the matrices led to establish a relationship between the matrix properties and its performance.

The measurements were analysed for the dependence of the performance on the molar extinction coefficients, geometry of matrix, substitution pattern within the matrix molecule and the functional groups at the carboxy carbon. As expected, molar extinction coefficient was found to be essential parameter for the matrix behaviour. Nevertheless, it was found that the parameter was not the sole deciding factor. Matrix **12** was well performing matrix with exceptionally small molar extinction coefficients. Many matrices like **2**, **7**, **8**, **18**, **32** have large extinction coefficients at 337 and 355 nm, still they performed poorly.

The analysis revealed importance of geometry and substitution pattern as performance improving factor. The para-biphenyl connectivity with para substitution that has minimum steric hindrance result in to better performance. A slight change in this, for example, meta-biphenyl connectivity or ortho-substitution lead to increased steric hindrance. This subsequently change the conformation of the molecule, which finally leads to diminished performance.

The number of hydrogen bonds formed also plays vital role in the matrix performance. The performance diminished as the amide hydrogens of primary amide were relaced by mono-ethyl and di-methyl substituents. This replacement led to decreased possibility to form the hydrogen bonds and resulted into the poor performance. The order of decreasing performance was primary amide> *N*-ethylamide > N, N-dimethylamide empasising the importance of H-bonding.

X-ray single crystal structures were obtained for some selected matrices. It was observed that in the best performing matrices **10** and **22** the H-bonding pattern is identical. Both the matrices show presence of H-bonds between R-C≡N|···H-NR_2 and R_2N-H· · ·|O=CR_2 functionalities, such that each molecule of **10** and **22** is connected to two adjacent matrix molecules by hydrogen bonds. If one of these two types of H-bonds was missing, (for example matrix **11** or **43** performance was found to be diminished. Thus along with the number of H-bonds possible, the pattern of H-bonding was also observed to play some role.

For large number of structural features and the physical properties of the matrices, statistical analysis was carried out to get some insight about the trends and correlations among the parameters. Principal component analysis revealed some interesting trends. For example, molar extinction coefficients and performances (relative mean S/N ratios) at 337 and 355 nm lasers were positively correlated as expected. The performance was negatively correlated to the logP, melting points and molecular weights. The number of exchangeable protons and pKa values were positively correlated to the performance parameters. H-bond donor atoms and laser powers used with the two different lasers were orthogonal to the relative mean S/N ratios implying no correlation between these parameters.

The identification of MALDI matrix is normally focused on the in depth evaluation and optimisation of experimental conditions for a single matrix compound. In the process detailed description of the matrix structure and its relationship with its performance remains unnoticed. To the best of author's knowledge, this was the first attempt to address the issue of identification of matrices and its structure-performance relationships with the medicinal chemistry approach. It was also the first time to incorporate PCA in the process of finding additional parameters that affect the performance of the matrix.

Lastly, it is worth mentioning here that there were several synthesised compounds which performed well as matrices for commonly used standard analytes like peptides from BSA digest and lipids from BTLE. The pyrrole derivatives were especially good for the detection of peptides in positive ion mode on 337 nm laser. These matrices were tried with different experimental conditions, nevertheless, they need more efforts in the optimisation to improve their performance.

6

Experimental Section

6.1 Experimental Section

6.1.1 General Information

Mass Spectrometry

Mass spectra were acquired on an UltrafleXtreme (355 nm Smartbeam II laser, modulated beam profile, 2000 Hz repetition rate, pressure ~ 3 x 10^{-7} hPa, Bruker Daltronics, Germany) and on an Ultraflex-I (337 nm nitrogen laser, 25 Hz repetition rate, pressure 2 x 10^{-7} hPa, Bruker Daltonics, Germany). For automated lipid measurements the AutoXecute function of the flexcontrol 3.4 software was used. Spectra were recorded in the mass range of *m/z* 400 to 2000 Da with a low mass gate 370 Da for negative ion mode. Only the spectra for which the software defined signal intensity level was above medium and with higher resolution than 5000 were accumulated. In case of 100 subsequently failed judgments the measurement was aborted. In sum 4000 laser shots were accumulated per spot by random walk with 25 shots and 50 shots at raster spot on Ultraflex-I and UltrafleXtreme respectively. The laser power was set individually for each matrix to get the best signal to noise ratio and peak intensity. The MS was calibrated internally using a theoretical mass list of detected lipids. Peak quality parameters were signal to noise threshold = 5 and peak width = 0.1 Da at 80% peak height. Spectra were analyzed with flexAnalysis 3.4 software (Bruker Daltonics).

Sample and matrix preparation for MALDI MS

1 µL of diluted (0.5 mg/mL) brain total lipid extract (Avanti Polar Lipids Inc, Alabaster, USA) per spot was transferred to a MALDI target plate (MTP 384 target plate polished steel BC, Bruker Daltonics, Bremen, Germany). The spots were air-dried and coated with 1 µL of matrix solution (5 replicates). If not stated differently, the biphenyl derivatives were used as 5 mg/mL solutions in acetonitrile/water (1:1, v/v) Acetonitrile was from Fisher Scientific (Acetonitrile Optima UHPLCMS, Leicestershire, UK). 9-Aminoacridine (Sigma-Aldrich, Steinheim, Germany) (5 mg/ml) was used as reference with the solvent system that was reported to achieve the best performance (2-propanol/acetonitrile 2:3, v/v)

Physical Properties Calculation

The physical properties pKa, logP, number of H-bond donors and number of H-bond acceptors were calculated on Instant JChem version 15.1.19.0 (ChemAxon, Budapest, Hungary).

Statistical Analysis

Data representation was done using Microsoft Excel (Microsoft Inc., Redmond, USA). S/N ratios and relative mean S/N ratios were analyzed separately for each matrix for different sulfatides against various calculated and/or experimentally determined physical properties and/or parameters. 59 biphenyl derivatives of HCCA were evaluated as MALDI-MS matrix compounds. The relative mean S/N value (RMSN) is calculated by dividing the sum of the relative S/N values through the number 14 of assigned sulfatides.

$$\text{Relative mean S/N ratio} = \frac{1}{14} \sum_{i=1}^{14} \frac{(\text{S/N matrix sulfatide species})_i}{(\text{S/N 9-AA sulfatide species})_i} \qquad (6.1)$$

NMR spectroscopy

The ^{1}H NMR (500 MHz) and ^{13}C NMR (125 MHz) were recorded at 298K with Bruker Avance DRX 500 instrument and using DMSO-d6 as solvent. DMSO-d6 for NMR were purchased from Carl-Roth, Germany or Deutero, Germany. The chemical shifts are expressed in ppm relative to TMS as an internal standard and the ^{1}H and ^{13}C are expressed with subscript *ole* for olefinic, *q* for quartenary and *ar* for aromatic protons and/or carbons.

Melting Points

Melting Points were determined with the help of Büchi melting point B-545 instrument (Büchi labortechnik AG, Flawil, Switzerland)

Materials and Reagents

All commercially available chemicals for synthesis were obtained from Sigma Aldrich GmbH (Taufkirchen, Germany) and Acros Organics (Fisher Scientific GmbH, Nidderau, Germany) and were used without any further purification. The reaction progress was monitored by TLC using cyclohexane/ethyl acetate mixtures. For analysis throughout this study, deionized water (Milli-Q, EVOQUA, Ultraclear-UV, Germany) was used and solvents (acetonitrile, acetone, chloroform, methanol) were generally obtained from Carl Roth, Karlsruhe, Germany in the highest available purity. The matrix 9-AA was purchased from Sigma Aldrich GmbH (Taufkirchen, Germany). Brain total lipid extract was purchased from Avanti Polar Lipids Inc. (Alabaster, Alabama, USA). According to the manufacturer's information, the extract contained 12.6% PC and 33.1% PE. It was stored as 5 mg/mL solution in chloroform/methanol (1:1, v/v) at −20 °C and diluted by addition of appropriate volumes of chloroform and methanol before preparation to obtain 0.5 mg/mL in chloroform/methanol (1:3, v/v)

Principal Component Analysis (PCA) and Cluster analysis (CA)

PCA and hierarchical cluster analysis performed with complete linkage method were carried out for all matrices using Minitab 17.2.1.0 (Minitab Ltd., Coventry, UK) and Origin 10.1 (OriginLab Corporation, Northampton, USA). For data pretreatment all values were transformed into the standard-score (z-values). The physical properties chosen for this analysis were logP, molecular weight, pKa, melting point, number of H-bond donors and acceptors, absorption values at 355nm and 337nm, number of atoms on the amide head group, biphenyl connectivity, substitution position on the aromatic ring, laser power used with the 337 nm and 355 nm lasers, number of exchangeable protons, free electron pairs on the amide head group whereas the parameters for matrix performance were relative mean S/N ratios for sulfatide using mass spectometers with 355 nm and 337 nm lasers. The values assigned for biphenyl connectivity were 2 = Scaffold a, 1 = Scaffold b and 0 = Scaffold c; and for substitution position 6 = no substituent, 4 = para-position, 3 = meta-position, and 0 = ortho-position.

MALDI-MS measurement parameters

For Ultraflex-I Spectra were recorded using negative ion reflector mode at ion source voltages IS1 20.00 kV and IS2 17.65 kV. The lens voltage is 6.00 kV. Reflector voltages were 21.00 kV for reflector 1 and 11.25 kV for reflector 2. For the instrument UltrafleXtreme voltages were 20.00 kV for IS1 and IS2 for 17.90 kV. The lens voltage was 7.50 kV. Reflector 1 voltage was 21.10 kV and reflector 2 voltage 10.85 kV.

X-ray Crystallography

Single compound crystals were grown at room temperature using a saturated acetone and water (7:3, v/v) stock solution. The crystallographic measurements were carried out on a Bruker APEX-II Quazar (matrices 19 and 45), Bruker APEX-II CCD (matrix 10) or a BRUKER APEX diffractometer (matrices 11, 22, 26) by Mr. Rominger, Heidelberg University, Germany. Intensities were corrected for Lorentz and polarisation effects, an empirical absorption corrections were applied using SADABS [1] based on the Laue symmetry of the reciprocal space. Structures were refined against F2 with a Full-matrix least-squares algorithm using the SHELXL-2014/7 software [2]. Crystal structure visualization and crystal packing calculation was done using Mercury 3.7 (build RC1), (Cambridge Crystallographic Data Centre, UK). CCDC 1484580 (matrix 10), 1484581 (matrix 11), 1484582 (matrix 19), 1484583 (matrix 22), 1484584 (matrix 26), 1484585 (matrix 45) contain the supplementary crystallographic data for this paper. These data can be obtained free of charge from The Cambridge Crystallographic Data Centre via www.ccdc.cam.ac.uk/data_request/cif.

General Procedure for the Synthesis of Cinnamic Acid Derivatives

A mixture of an aromatic aldehyde (1.0 eq., Sigma Aldrich, Steinheim, Germany), 2-cyanoacetic acid/amides (2.0 eq., Sigma Aldrich, Steinheim, Germany) and ammonium acetate (0.2 eq.) was dissolved in toluene (or ethanol) and refluxed until complete conversion (2-6 hrs) was shown by thin-layer chromatography . The product was purified by recrystallization (2-3 times) from acetone/water mixtures or by flash chromatography (Ethylacetate/Cyclohexane) and characterized by ^{1}H, ^{13}C-NMR, UV-Vis and MALDI-MS.

UV-Vis Measurements

Solution phase absorption spectra of the matrices were acquired with a Cary 60 UV-Vis spectrophotometer (Agilent Technologies, Santa Clara, USA) using 1 cm optical quartz cuvettes (Hellma Analytics, Mühlheim, Germany). Extinction coefficients of compounds were determined in methanol (HPLC grade, Carl Roth GmbH & Co.KG)/water (4:1, v/v). Solid state UV-Vis measurements were carried out on a home-built instrument (N. Hellmuth, Aalen University of Applied Sciences, Germany) using Multispec© View software (Purdue Research Foundation, West Lafayette, USA): In briefly, a fiberlight deuterium/wolfram lamp (Heraeus, Hanau, Germany) light source was connected to a 400 µm diameter optical quartz fiber. The size of the fiber outlet was 0.5 mm. 1-3 µL matrix solution (2-5 mg/mL in ethanol) was transferred on the fiber outlet and dried. For the measurements the fiber outlet was fixed with a removable screw fitting. The light path between the sample and the detector was focused with two quartz lenses. The first lens (focal distance 30 mm) was the collimating lens; the second lens (focal distance 20 mm) was the convergence lens. The collimated light beam was collected with a second quartz fiber with a diameter of 400 µm which was attached to the MMS-UV/VIS (Carl Zeiss, Jena, Germany, 190-720 nm wavelength range) detector.

6.1.2 Characterisation:

6.1.2.1 Phenylcinnamic Acid Derivatives

(2*E*)-3-[4-(2H-1,3-benzodioxol-5-yl)phenyl]-2-cyanoprop-2-enamide (1)

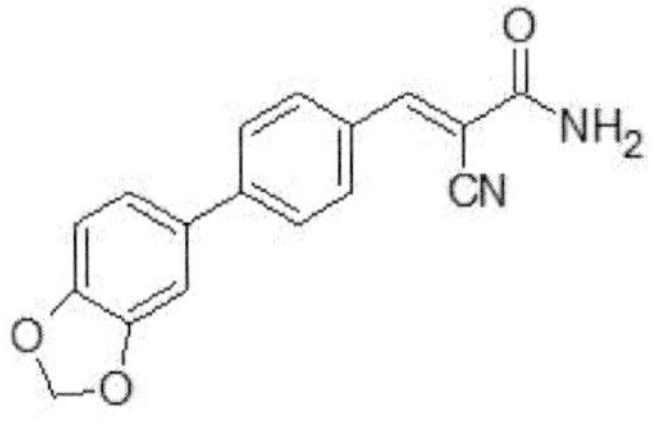

Pale yellow crystals, 52% yield, Melting Point = 217.7 °C UV-Vis (MeOH: H_2O, 8:2,v/v):λ_{max} = 323 nm MALDI-MS: $[M+H]^+$ = 293.35 Da ^{1}H-NMR (500 MHz, DMSO-d6): δ [ppm] = 8.20 (s, 1H, CHole), 8.00 (d, $^3J_{H,H}$ = 8.20 Hz, 2H, 2 x CHar), 7.92 (s, br, 1H, NH), 7.84 (d, $^3J_{H,H}$ = 8.20 Hz, 2H, 2 x CHar), 7.77 (s, br, 1H, NH), 7.39 (s, 1H, CHar), 7.30 (d, $^3J_{H,H}$ = 6.90 Hz, 1H, CHar), 7.05 (d,$^3J_{H,H}$ = 8.20 Hz, 1H, CHar), 6.10 (s, 2H, CH_2). ^{13}C-NMR (125 MHz, DMSO-d6): δ [ppm] = 162.7 (1C, CO), 149.9 (1C, CHar), 148.1 (1C, CHole), 147.6 (1C, Cq), 130.6 (2C, 2 x CHar), 130.3 (1C, Cq), 130.0 (1C, CHar), 126.9 (1C, Cq), 126.8 (2C, 2 x CHar), 121.1 (1C, CHar), 120.8 (1C, CHar), 108.7 (1C, CHar), 107.1 (1C, CHar), 105.6 (1C, CN), 101.3 (1C, CH_2).

(2*E*)-3-([1,1'-biphenyl]-4-yl)-N-carbamoyl-2-cyanoprop-2-enamide (2)

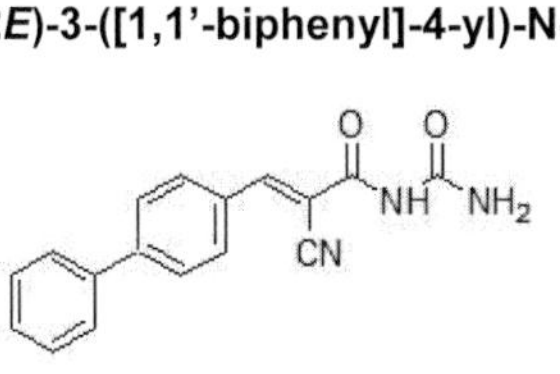

Pale yellow crystals, 84% yield, Melting Point = 245.4°C UV-Vis (MeOH: H_2O, 8:2,v/v):λ_{max} = 342 nm MALDI-MS: $[M+H]^+$ = 292.40 Da

^{1}H-NMR (500 MHz, DMSO-d6): δ [ppm] = 10.63 (s, br, 1H, NH), 8.41 (s, 1H, CHole), 8.07 (d, $^3J_{H,H}$ = 7.90 Hz, 2H, 2 x CHar), 7.94 (d, $^3J_{H,H}$ = 7.30 Hz, 2H, 2 x CHar), 7.80 (d, $^3J_{H,H}$ = 6.90 Hz, 3H, 2 x CHar), 7.70 (s, br, 1H, NH), 7.54-7.51 (m, 3H, 2 x CHar, NH), 7.46-7.43 (m, 1H, CHar). ^{13}C-NMR (125 MHz, DMSO-d6): δ [ppm] = 163.6 (1C, CO), 153.2 (1C, CO), 151.6 (1C, CHole), 144.2 (1C, Cq), 138.4 (1C, Cq), 131.1 (2C, 2 x CHar), 130.4 (1C, Cq), 129.1 (2C, 2 x CHar), 128.5 (1C, CHar), 127.3 (2C, 2 x CHar), 126.9 (2C, 2 x CHar), 115.6 (1C, Cq), 105.3 (1C, CN).

(2*E*)-3-([1,1'-biphenyl]-4-yl)-N-carbamoyl-2-cyanoprop-2-enamide (3)

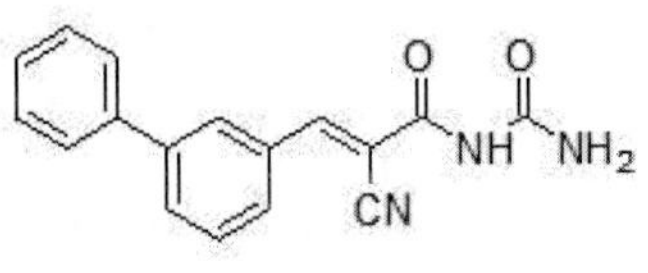

White crystals, 76% yield, Melting Point = 208.7 °C UV-Vis (MeOH: H_2O, 8:2,v/v):λ_{max} = 307 nm MALDI-MS: $[M+H]^+$ = 292.37 Da ^{1}H-NMR (500 MHz, DMSO-d6): δ [ppm] = 10.64 (s, br, 1H, NH), 8.47 (s, 1H, CHole), 8.25 (s, 1H, CHar), 7.95-7.92 (m, 2H, 2 x CHar), 7.72-7.69 (m, 4H, 3 x CHar, NH), 7.55-7.52 (m, 3H, 2 x CHar, NH), 7.45-7.42 (m, 1H, CHar). ^{13}C-NMR (125 MHz, DMSO-d6): δ [ppm] = 163.4 (1C, CO), 153.1 (1C, CO), 152.2 (1C, CHole), 140.9 (1C, Cq), 138.8 (1C, Cq), 132.1 (1C, Cq), 131.0 (1C, CHar), 129.9 (1C, CHar), 129.1 (2C, 2 x CHar), 129.0 (1C, CHar), 128.4 (1C, CHar), 128.0 (1C, CHar), 126.6 (2C, 2 x CHar), 115.5 (1C, Cq), 106.4 (1C, CN).

(2*E*)-2-cyano-3-(3'-hydroxy[1,1'-biphenyl]-3-yl)prop-2-enamide (4)

Pale yellow solid, 84% yield, Melting Point = 204.0 °C UV-Vis (MeOH: H_2O, 8:2,v/v):λ_{max} = 299 nm MALDI-MS: $[M+H]^+$ = 265.45 Da ^{1}H-NMR (500 MHz, DMSO-d6): δ [ppm] = 9.65 (s, br, 1H, OH), 8.27 (s, 1H, CHar), 8.18 (s, 1H, CHole), 7.95 (s,

1H, br, NH), 7.90 (d, $^3J_{H,H}$ = 7.30 Hz,1H, CHar), 7.81 (d,$^3J_{H,H}$ = 6.90 Hz, 2H, CHar, NH), 7.66-7.63 (m, 1H, CHar), 7.30 (t,$^3J_{H,H}$ = 7.90 Hz, 1H, CHar), 7.11 (d, $^3J_{H,H}$ = 7.90 Hz, 1H, CHar), 7.06 (s, 1H, CHar), 6.82 (d,$^3J_{H,H}$ = 7.60 Hz, 1H, CHar). ^{13}C-NMR (125 MHz, DMSO-d6): δ [ppm] = 162.6 (1C, CO), 157.8 (1C, Cq), 150.5 (1C, CHole), 141.0 (1C, Cq), 140.4 (1C, Cq), 132.4 (1C, Cq), 130.3 (1C, CHar), 130.0 (1C, CHar), 129.7 (1C, CHar), 128.8 (1C, CHar), 127.9 (1C, CHar), 117.4 (1C, CHar), 116.4 (1C, Cq), 114.9 (1C, CHar), 113.5 (1C, CHar), 107.0 (1C, CN).

(2*E*)-2-cyano-3-(4'-hydroxy[1,1'-biphenyl]-2-yl)prop-2-enamide (5)

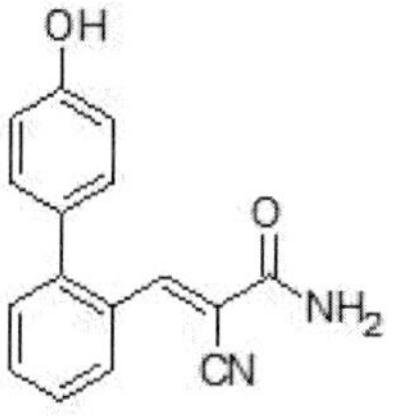

Yellow solid, 84% yield, Melting Point = 239.5 °C UV-Vis (MeOH: H_2O, 8:2,v/v):λ_{max} = 338 nm MALDI-MS: $[M+H]^+$ = 265.48 Da ^{1}H-NMR (500 MHz, DMSO-d6): δ [ppm] = 9.76 (s, br, 1H, OH), 7.98 (s, 2H, CHole, CHar), 7.81 (s, 1H, br, NH), 7.70 (s, br, 1H, NH), 7.62-7.59 (m, 1H, CHar), 7.53-7.49 (m, 2H, 2 x CHar), 7.13 (d, $^3J_{H,H}$ = 8.20 Hz, 2H, 2 x CHar), 6.87 (d, $^3J_{H,H}$ = 8.20 Hz, 2H, 2 x CHar) . ^{13}C-NMR (125 MHz, DMSO-d6): δ [ppm] = 162.2 (1C, CO), 157.5 (1C, Cq), 151.3 (1C, CHole),143.0 (1C, Cq), 131.6 (1C, CHar), 131.0 (2C, 2 x CHar), 130.0 (1C, CHar), 129.8 (1C, Cq), 129.2 (1C, Cq), 128.4 (1C, CHar), 126.9 (1C, CHar), 115.3 (2C,2 x CHar), 107.9 (1C, CN).

(2*E*)-2-cyano-3-(4'-methoxy[1,1'-biphenyl]-4-yl)prop-2-enamide (6)

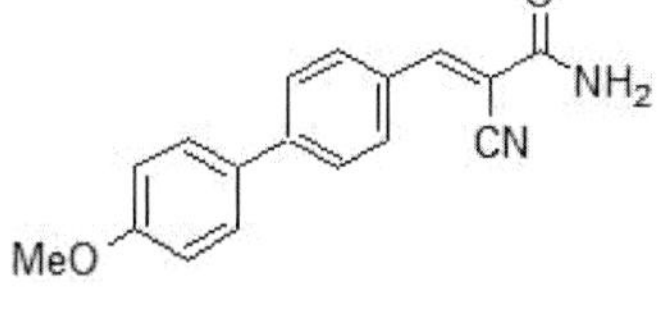

Pale yellow crystals, 73% yield, Melting Point = 215.2 °C UV-Vis (MeOH: H_2O, 8:2,v/v):λ_{max} = 341 nm MALDI-MS: $[M+H]^+$ = 279.56 Da ^{1}H-NMR (500 MHz, DMSO-d6): δ [ppm] = 8.20 (s, 1H, CHole), 8.02 (d, 3J H,H = 8.20 Hz, 2H, 2 x CHar), 7.92 (s, br, 1H, NH), 7.86 (d, $^3J_{H,H}$ = 8.20 Hz,

2H, 2 x CHar), 7.75 (d, $^3J_{H,H}$ = 8.50 Hz, 3H, 2 x CHar, NH), 7.07 (d, $^3J_{H,H}$ = 8.80 Hz, 2H, 2 x CHar), 3.82 (s, 3H, CH_3). ^{13}C-NMR (125 MHz, DMSO-d6): δ [ppm] = 162.7 (1C, CO), 159.4 (1C, Cq), 145.8 (1C, CHole), 142.0 (1C, Cq), 130.8 (1C, Cq), 130.3 (2C, 2 x CHar), 127.9 (2C, 2 x CHar), 126.5 (2C, 2 x CHar), 116.6 (1C, Cq), 114.4 (2C, 2 x CHar), 105.4 (1C, CN), 55.1 (1C, CH_3).

(2*E*)-2-cyano-3-(3'-methoxy[1,1'-biphenyl]-4-yl)prop-2-enamide (7)

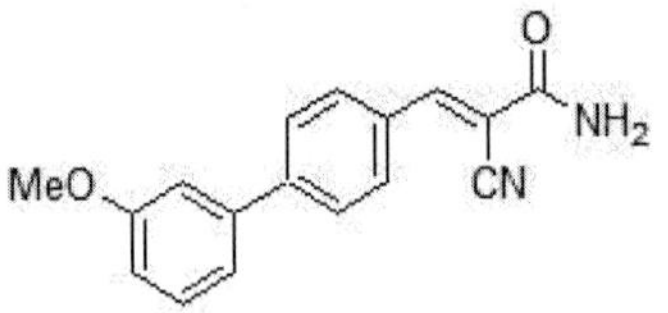

Pale yellow crystals, 94% yield, Melting Point = 165.6 °C UV-Vis (MeOH: H_2O, 8:2,v/v):λ_{max} = 335 nm MALDI-MS: $[M+H]^+$ = 279.53 Da ^{1}H-NMR (500 MHz, DMSO-d6): δ [ppm] = 8.22 (s, 1H, CHole), 8.04 (d, $^3J_{H,H}$ = 8.20 Hz, 2H, 2 x CHar), 7.94 (s, br, 1H, NH), 7.91 (d,$^3J_{H,H}$ = 8.20 Hz, 2H, 2 x CHar), 7.79 (s, br, 1H, NH), 7.44-7.41 (m, 1H, CHar), 7.36 (d, $^3J_{H,H}$ = 7.30 Hz, 1H, CHar), 7.32 (s, 1H, CHar), 7.01 (d, $^3J_{H,H}$ = 8.50 Hz, 1H, CHar), 3.84 (s, 3H, CH_3). ^{13}C-NMR (125 MHz, DMSO-d6): δ [ppm] = 162.7 (1C, CO), 159.7 (1C, Cq), 149.9 (1C, CHole), 143.5 (1C, Cq), 140.1 (1C, Cq), 130.9 (1C, Cq), 130.6 (2C, 2 x CHar), 130.1 (1C, CHar), 127.4 (2C, 2 x CHar), 119.1 (1C, CHar), 116.5 (1C, Cq), 114.1 (1C, CHar), 112.2 (1C, CHar), 106.1 (1C, CN), 55.1 (1C, CH_3).

(2*E*)-2-cyano-3-(2'-methoxy[1,1'-biphenyl]-4-yl)prop-2-enamide (8)

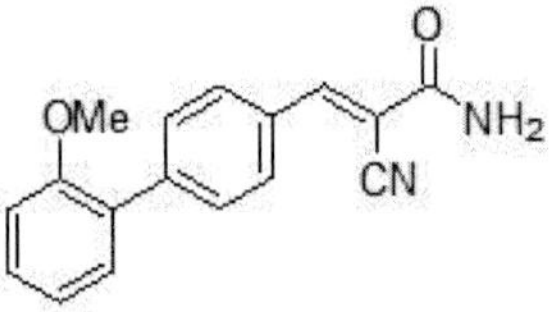

Pale yellow crystals, 87% yield, Melting Point = 194.1 °C UV-Vis (MeOH: H_2O, 8:2,v/v):λ_{max} = 339 nm MALDI-MS: $[M+H]^+$ = 279.49 Da ^{1}H-NMR (500 MHz, DMSO-d6): δ [ppm] = 8.22 (s, 1H, CHole), 7.99 (d, $^3J_{H,H}$ = 8.20 Hz, 2H, 2 x CHar), 7.95 (s, br,

1H, NH), 7.80 (s, br, 1H, NH), 7.70 (d,$^3J_{H,H}$ = 8.20 Hz, 2H, 2 x CHar), 7.41-7.38 (m, 2H, 2 x CHar), 7.17 (d, $^3J_{H,H}$ = 8.20 Hz, 1H, CHar), 7.08 (t, $^3J_{H,H}$ = 7.10 Hz, 1H, CHar), 3.81 (s, 3H, CH_3). ^{13}C-NMR (125 MHz, DMSO-d6): δ [ppm] = 162.7 (1C, CO), 156.1 (1C, Cq), 150.1 (1C, CHole), 142.1 (1C, Cq), 130.3 (1C, CHar), 130.2 (1C, Cq), 129.9 (2C, 2 x CHar), 129.7 (2C, 2 x CHar), 129.6 (1C, CHar), 128.3 (1C, Cq), 120.8 (1C, CHar), 116.5 (1C, Cq), 111.8 (1C, CHar), 105.8 (1C, CN), 55.5 (1C, CH_3).

Methyl-3'-[(1*E*)-3-amino-2-cyano-3-oxoprop-1-en-1-yl][1,1'-biphenyl]-4-carboxylate (9)

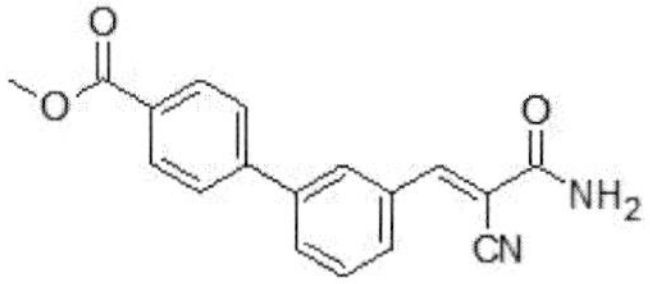

Light brown solid, 16% yield, Melting Point = 208.9 °C UV-Vis (MeOH: H_2O, 8:2,v/v):λ_{max} = 276 nm MALDI-MS: $[M+H]^+$ = 307.37 Da ^{1}H-NMR (500 MHz, DMSO-d6): δ [ppm] = 8.31 (s, 1H, CHar), 8.28 (s, 1H, CHole), 8.09 (d,$^3J_{H,H}$ = 7.50 Hz, 2H, 2 x CHar), 8.00 (d, $^3J_{H,H}$ = 7.50 Hz, 1H, CHar), 7.96 (d, $^3J_{H,H}$ = 7.50 Hz, 1H, CHar), 7.95 (s, br, 1H, NH), 7.88 (d,$^3J_{H,H}$ = 7.50 Hz, 2H, 2 x CHar), 7.82 (s, br, 1H, NH), 7.71 (t, $^3J_{H,H}$ = 7.50 Hz, 1H, CHar), 3.89 (s, 3H, CH_3) . ^{13}C-NMR (125 MHz, DMSO-d6): δ [ppm] = 165.8 (1C, CO), 150.4 (1C, CHole), 130.6 (1C, CHar), 130.0 (1C, CHar), 129.9 (2C, 2 x CHar), 129.2 (1C, CHar), 128.9 (1C, Cq), 128.7 (1C, CHar), 127.0 (2C, 2 x CHar), 116.4 (1C, Cq), 107.4 (1C, CN), 52.1 (1C, CH_3).

(2*E*)-3-([1,1'-biphenyl]-3-yl)-2-cyanoprop-2-enamide (11)

White solid, 55% yield, Melting Point = 253.4 °C

UV-Vis (MeOH: H_2O, 8:2,v/v):λ_{max} = 300 nm MALDI-MS: $[M+H]^+$ = 249.27 Da ^{1}H-NMR (500 MHz, DMSO-d6): δ [ppm] = 8.29 (s, 1H, CHar), 8.25 (s, 1H, CHole), 7.96 (s, 1H, br,

NH), 7.93 (d, $^3J_{H,H}$ = 7.90 Hz, 1H, CHar), 7.89 (d, $^3J_{H,H}$ = 7.85 Hz, 1H, CHar), 7.81 (s, br, 1H, NH), 7.72 (d, $^3J_{H,H}$ = 7.45 Hz, 2H, 2 x CHar), 7.67 (t, $^3J_{H,H}$ = 7.75 Hz, 1H, CHar), 7.52 (t, $^3J_{H,H}$ = 7.60 Hz, 2H,2 x CHar), 7.43 (t, $^3J_{H,H}$ = 7.20 Hz, 1H, CHar). ^{13}C-NMR (125 MHz, DMSO-d6): δ [ppm] = 162.5 (1C, CO), 150.5 (1C, CHole), 140.9 (1C, Cq), 139.0 (1C, Cq), 132.5 (1C, Cq), 130.4 (1C, CHar), 129.8 (1C, CHar), 129.0 (2C, 2 x CHar), 128.7 (1C, CHar), 128.1 (1C, CHar), 127.9 (1C, CHar), 126.6 (2C, 2 x CHar), 116.4 (1C, Cq), 107.1 (1C, CN).

(2*E*)-3-([1,1'-biphenyl]-2-yl)-2-cyanoprop-2-enamide (12)

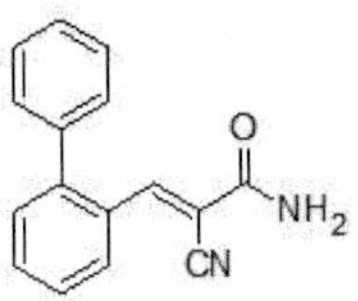

White solid, 84% yield, Melting Point = 139.9 °C UV-Vis (MeOH: H_2O, 8:2,v/v):λ_{max} = 301 nm MALDI-MS: $[M+H]^+$ = 249.23 Da ^{1}H-NMR (500 MHz, DMSO-d6): δ [ppm] = 8.03 (d, $^3J_{H,H}$ = 7.60 Hz, 1H, CHar), 7.98 (s, 1H, CHole), 7.82 (s, 1H, br, NH), 7.72 (s, 1H, br, NH), 7.68-7.65 (m, 1H, CHar), 7.61-7.58 (m, 1H, CHar), 7.56 (d, $^3J_{H,H}$ = 7.30 Hz, 1H, CHar), 7.53-7.47 (m, 3H, 3 x CHar), 7.34 (d, $^3J_{H,H}$ = 7.30 Hz, 2H, 2 x CHar) . ^{13}C-NMR (125 MHz, DMSO-d6): δ [ppm] = 162.0 (1C, CO), 150.9 (1C, CHole),142.8 (1C, Cq), 138.7 (1C, Cq), 131.6(1C, Cq), 130.3 (1C, Cq), 130.1 (1C, CHar), 129.7 (1C, CHar), 128.5 (2C, 2 x CHar), 128.0 (1C, CHar), 127.8 (1C, CHar), 116.2 (1C, Cq), 108.6 (1C, CN).

(2*E*)-2-cyano-3-(9*H*-fluoren-2-yl)prop-2-enamide (13)

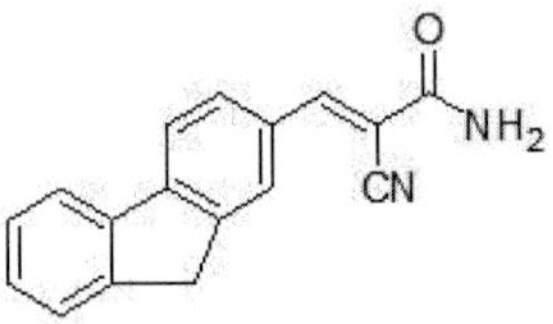

Yellow crystals, 82% yield, Melting Point = 195.6 °C UV-Vis (MeOH: H_2O, 8:2,v/v):λ_{max} = 350 nm MALDI-MS: $[M+H]^+$ = 261.33 Da ^{1}H-NMR (500 MHz, DMSO-d6): δ [ppm] = 8.25 (s, 1H, CHar), 8.21 (s, 1H, CHole), 8.11 (d, $^3J_{H,H}$ = 8.20 Hz, 1H, CHar), 8.01 (dd, $^3J_{H,H}$ = 14.20 Hz, $^3J_{H,H}$ = 7.90

Hz, 2H, 2 x CHar), 7.92 (s, br, 1H, NH), 7.77 (s, br, 1H, NH), 7.65 (d, $^3J_{H,H}$ = 6.90 Hz, 1H, CHar), 7.47-7.40 (m, 2H, 2 x CHar), 4.04 (s, 2H, CH_2). ^{13}C-NMR (125 MHz, DMSO-d6): δ [ppm] = 162.9 (1C, CO), 150.7 (1C, CHole), 145.2 (1C, Cq), 144.2 (1C, Cq), 143.6 (1C, Cq), 139.8 (1C, Cq), 130.2 (1C, Cq), 129.7 (1C, CHar), 128.2 (1C, CHar), 127.0 (1C, CHar), 126.4 (1C, CHar), 125.3 (1C, CHar), 121.0 (1C, CHar), 120.6 (1C, CHar), 116.8 (1C, Cq), 104.9 (1C, CN), 36.3 (1C, CH_2).

(2*E*)-2-cyano-3-(4'-fluoro[1,1'-biphenyl]-4-yl)prop-2-enamide (14)

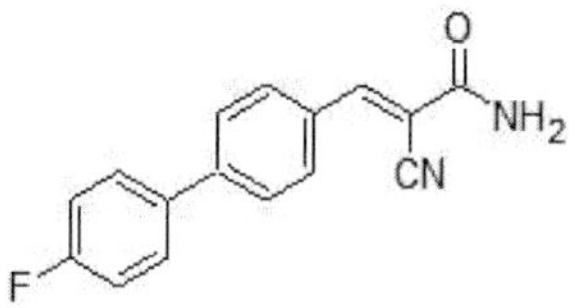

Yellow crystals, 53% yield, Melting Point = 200.9 °C UV-Vis (MeOH: H_2O, 8:2,v/v): λ_{max} = 334 nm MALDI-MS: $[M+H]^+$ = 267.31 Da ^{1}H-NMR (500 MHz, DMSO-d6): δ [ppm] = 8.22 (s, 1H, CHole), 8.04 (d, $^3J_{H,H}$ = 7.95 Hz, 2H, 2 x CHar), 7.94 (s, br, 1H, NH), 7.89 (d, $^3J_{H,H}$ = 8.30 Hz, 2H, 2 x CHar), 7.86-7.83 (m, 2H, 2 x CHar), 7.79 (s, br, 1H, NH), 7.35 (t, $^3J_{H,H}$ = 8.90 Hz, 2H, 2 x CHar). ^{13}C-NMR (125 MHz, DMSO-d6): δ [ppm] = 162.6 (1C, CO),162.5 (d, $^1J_{C,F}$ = 240 Hz, 1C, CF), 149.8 (1C, CHole), 142.5 (1C, Cq), 135.6 (1C, Cq), 135.0 (1C, Cq), 130.7 (2C, 2 x CHar), 129.0 (1C, Cq), 128.9 (2C, CHar), 127.2 (2C, 2 x CHar), 116.5 (1C, Cq), 115.8 (d, $^2J_{C,F}$ = 23 Hz, 2C, 2 x CHar), 106.1 (1C, CN).

(2*E*)-2-cyano-3-(3'-fluoro[1,1'-biphenyl]-3-yl)prop-2-enamide (15)

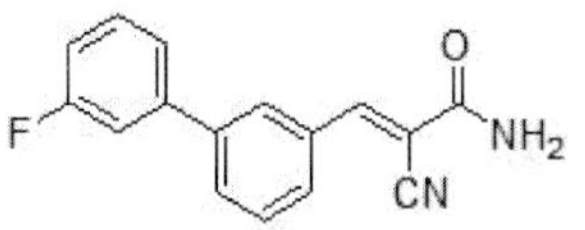

White crystals, 52% yield, Melting Point = 120.6 °C UV-Vis (MeOH: H_2O, 8:2,v/v): λ_{max} = 298 nm MALDI-MS: $[M+H]^+$ = 267.31 Da ^{1}H-NMR (500 MHz, DMSO-d6): δ [ppm] = 8.30 (s, 1H, CHole), 8.26 (s, 1H, CHar), 7.97 (d, $^3J_{H,H}$ = 7.30 Hz, 2H, 2 x CHar), 7.94 (d, $^3J_{H,H}$ = 6.90 Hz, 1H, CHar), 7.83 (s, br, 1H, NH), 7.68 (t, $^3J_{H,H}$ = 7.70 Hz, 1H, CHar), 7.57 (t, $^3J_{H,H}$ = 8.70 Hz, 3H, 2 x CHar, NH), 7.28-7.25 (m, 1H, CHar). ^{13}C-NMR (125 MHz, DMSO-d6): δ [ppm] = 165.2 (1C, CO), 151.0 (1C, CHole), 140.0 (1C, Cq), 133.2 (1C, Cq), 131.6 (1C, CHar), 131.0 (1C, CHar), 130.4 (1C, CHar), 129.6 (1C, CHar), 129.1 (1C, CHar), 126.8 (1C, Cq), 123.3 (1C, CHar), 116.9 (1C, Cq), 115.4 (d, $^2J_{C,F}$ = 23 Hz, 1C, CHar), 114.1 (1C, CHar), 107.8 (1C, CN).

(2*E*)-*N*-carbamoyl-2-cyano-3-(4'-methyl[1,1'-biphenyl]-4-yl)prop-2-enamide (16)

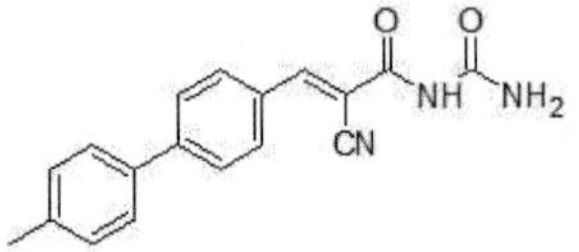

Yellow crystals, 87% yield, Melting Point = 248.3 °C UV-Vis (MeOH: H_2O, 8:2,v/v):λ_{max} = 300 nm MALDI-MS: $[M+H]^+$ = 306.41 Da ^{1}H-NMR (500 MHz, DMSO-d6): δ [ppm] = 10.61 (s, br, 1H, NH), 8.39 (s, 1H, CHole), 8.05 (d, $^3J_{H,H}$ = 8.50 Hz, 2H, 2 x CHar), 7.91 (d, $^3J_{H,H}$ = 8.80 Hz, 2H, 2 x CHar), 7.70 (d, $^3J_{H,H}$ = 7.60 Hz, 3H, 2 x CHar, NH), 7.51 (s, br, 1H, NH), 7.33 (d, $^3J_{H,H}$ = 6.90 Hz, 2H, 2 x CHar), 2.37 (s, 3H, CH3). ^{13}C-NMR (125 MHz, DMSO-d6): δ [ppm] = 163.6 (1C, CO), 153.2 (1C, CO), 151.6 (1C, CHole), 144.1 (1C, Cq), 138.2 (1C, Cq), 135.5 (1C, Cq), 131.1 (2C, 2 x CHar), 130.1 (1C, Cq), 129.6 (2C, 2 x CHar), 126.9 (2C, 2 x CHar), 126.7 (2C, 2 x CHar), 115.7 (1C, Cq), 105.0 (1C, CN), 20.6 (1C, CH_3).

(2*E*)-2-cyano-3-(2',4'-difluoro[1,1'-biphenyl]-4-yl)prop-2-enamide (17)

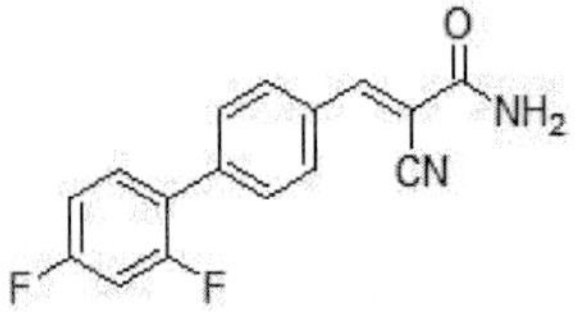

White crystals, 31% yield, Melting Point = 236.3 °C UV-Vis (MeOH: H_2O, 8:2,v/v):λ_{max} = 325 nm MALDI-MS: $[M+H]^+$ = 285.37 Da ^{1}H-NMR (500 MHz, DMSO-d6): δ [ppm] = 8.23 (s, 1H, CHole), 8.05 (d, $^3J_{H,H}$ = 8.20 Hz, 2H, 2 x CHar), 7.96 (s, br, 1H, NH), 7.82 (s, br, 1H, NH), 7.75 (d, $^3J_{H,H}$ = 8.20 Hz, 2H, 2 x CHar), 7.69 (td, $^3J_{H,F}$ = 9 Hz, 3J H,H = 6.60 Hz, 1H, CHar), 7.42 (ddd, $^3J_{H,F}$ = 11.3 Hz, $^3J_{H,H}$ = 9.10 Hz, $^4J_{H,H}$ = 2.7 Hz 1H, CHar), 7.25 (td, $^3J_{H,H}$ = 8.40 Hz, , $^4J_{H,F}$ = 7.0 Hz, $^3J_{H,H}$ = 2.40 Hz, 1H, CHar). ^{13}C-NMR (125 MHz, DMSO-d6): δ [ppm] = 163.1 (1C, CO), 162.7 (dd, $^1J_{C,F}$ = 249 Hz, $^3J_{C,F}$ = 12 Hz, 1C, CF), 159.7 ((dd, $^1J_{C,F}$ = 251 Hz, $^3J_{C,F}$ = 12 Hz, 1C, CF), 150.4 (1C, CHole), 138.4 (1C, Cq), 132.4 (dd, $^3J_{C,F}$ = 10 Hz, $^4J_{C,F}$ = 4.5 Hz, 1C, CHar), 131.8 (1C, Cq), 130.9 (2C, 2 x CHar), 130.0 (2C, 2 x CHar), 124.2 (1C, Cq), 116.4 (1C, Cq), 112.9 (d, , $^1J_{C,F}$ = 18.3 Hz, 1C, CHar), 107.3 (1C, CN), 105.2 (t, $^2J_{C,F}$ = 25 Hz, 1C, CHar).

(2*E*)-3-[4-(2*H*-1,3-benzodioxol-5-yl)phenyl]-2-cyanoprop-2-enoic acid (18)

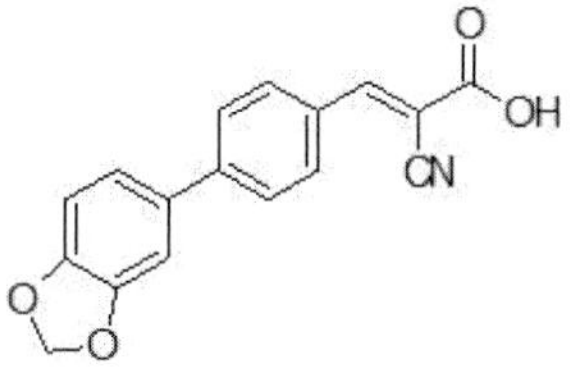

Yellow crystals, 61% yield, Melting Point = 274.0 °C UV-Vis (MeOH: H_2O, 8:2,v/v):λ_{max} = 343 nm MALDI-MS: $[M+H]^+$ = 293.35 Da ^{1}H-NMR (500 MHz, DMSO-d6): δ [ppm] = 8.35 (s, 1H, CHole), 8.10 (d, $^3J_{H,H}$ = 7.60 Hz, 2H, 2 x CHar), 7.86 (d, $^3J_{H,H}$ = 8.50 Hz, 2H, 2 x CHar), 7.41 (s, 1H, CHar), 7.32 (d,$^3J_{H,H}$ = 6.30 Hz, 1H, CHar), 7.05 (d, $^3J_{H,H}$ = 8.20 Hz, 1H, CHar), 6.10 (s, 2H, CH_2). ^{13}C-NMR (125 MHz, DMSO-d6): δ [ppm] = 163.3 (1C, CO), 153.6 (1C, Cq), 148.1 (1C, CHole), 131.3 (2C, 2 x CHar), 129.9 (1C, CHar), 126.8 (2C, 2 x CHar), 120.9 (1C, CHar), 108.7 (1C, CHar), 107.1 (1C, CN), 101.3 (1C, CH_2).

(2*E*)-3-([1,1'-biphenyl]-4-yl)-2-cyano-N,N-dimethylprop-2-enamide (19)

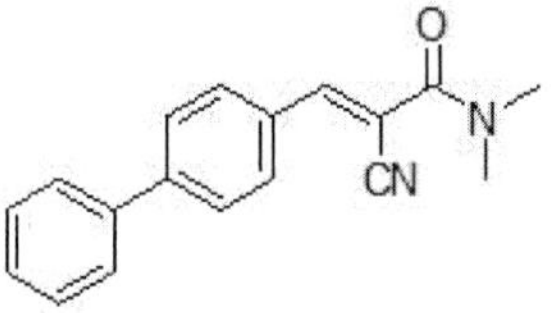

White crystals, 94% yield, Melting Point = 125.3 °C UV-Vis (MeOH: H_2O, 8:2,v/v):λ_{max} = 328 nm MALDI-MS: $[M+H]^+$ = 277.34 Da ^{1}H-NMR (500 MHz, DMSO-d6): δ [ppm] = 7.99 (d,$^3J_{H,H}$ = 8.20 Hz, 2H, 2 x CHar), 7.79 (s, 1H, CHole), 7.71 (d, $^3J_{H,H}$ = 7.30 Hz, 2H, 2 x CHar), 7.64 (d, $^3J_{H,H}$ = 7.30 Hz, 2H, 2 x CHar), 7.51-7.46 (m, 2H, 2 x CHar), 7.43-7.38 (m, 1H, CHar), 3.24 (s, 3H, CH_3), 3.09 (s, 3H, CH_3). ^{13}C-NMR (125 MHz, DMSO-d6): δ [ppm] = 163.0 (1C, CO), 149.1 (1C, CHole), 143.2 (1C, Cq), 138.6 (1C, Cq), 131.2 (1C, Cq), 130.1 (2C, 2 x CHar), 129.0 (2C, 2 x CHar), 128.3 (1C, CHar), 127.1 (2C, 2 x CHar), 126.8 (2C, 2 x CHar), 116.2 (1C, Cq), 105.5 (1C, CN),37.0 (1C, CH_3), 34.3 (1C, CH_3).

(2*E*)-2-cyano-3-(3'-hydroxy[1,1'-biphenyl]-3-yl)prop-2-enoic acid (20)

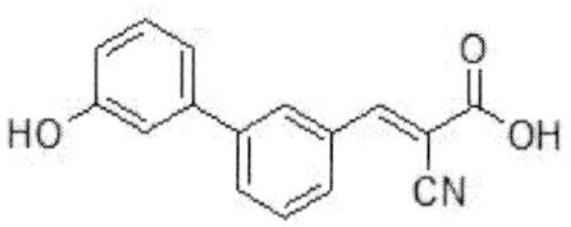

Yellow solid, 84% yield, Melting Point = 187.9 °C UV-Vis (MeOH: H_2O, 8:2,v/v):λ_{max} = 293 nm MALDI-MS: $[M+H]^+$ = 266.35 Da ^{1}H-NMR (500

MHz, DMSO-d6): δ [ppm] = 9.64 (s, br, 1H, OH), 8.44 (s, 1H, CHar), 8.29 (s, 1H, CHole), 8.01 (d, $^3J_{H,H}$ = 7.30 Hz, 1H, CHar), 7.85 (d, $^3J_{H,H}$ = 7.60 Hz, 1H, CHar), 7.66 (t, $^3J_{H,H}$ = 8.00 Hz, 1H, CHar), 7.32-7.28 (m, 1H, CHar), 7.12 (d,$^3J_{H,H}$ = 7.30 Hz, 1H, CHar), 7.07 (s, 1H, CHar), 6.82 (d, $^3J_{H,H}$ = 7.60 Hz, 1H, CHar). ^{13}C-NMR (125 MHz, DMSO-d6): δ [ppm] = 163.1 (1C, CO), 157.8 (1C, Cq), 154.3 (1C, CHole), 141.1 (1C, Cq), 140.3 (1C, Cq), 132.1 (1C, Cq), 131.0 (1C, CHar), 130.0 (1C, CHar), 129.8 (1C, CHar), 129.3 (1C, CHar), 128.6 (1C, CHar), 117.4 (1C, CHar), 116.1 (1C, Cq), 114.9 (1C, CHar), 113.5 (1C, CHar), 104.2 (1C, CN).

(2*E*)-2-cyano-3-(4'-hydroxy[1,1'-biphenyl]-2-yl)prop-2-enoic acid (21)

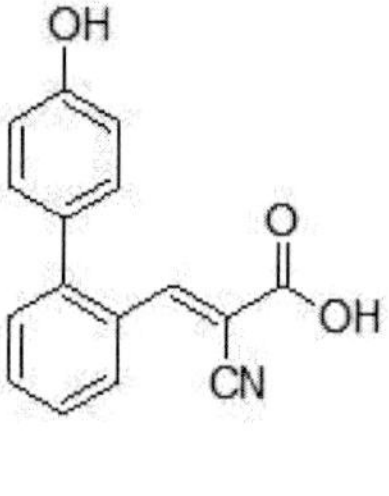

Pale yellow solid, 81% yield, Melting Point = 255.3 °C UV-Vis (MeOH: H_2O, 8:2,v/v):λ_{max} = 288 nm MALDI-MS: $[M+H]^+$ = 266.35 Da ^{1}H-NMR (500 MHz, DMSO-d6): δ [ppm] = 9.79 (s, br, 1H, OH), 8.11 (d, $^3J_{H,H}$ = 7.60 Hz, 1H, CHar), 8.07 (s, 1H, CHole), 7.67-7.64 (m, 1H, CHar), 7.57-7.52 (m, 2H, 2 x CHar), 7.16 (d, $^3J_{H,H}$ = 8.20 Hz, 2H, 2 x CHar), 6.88 (d, $^3J_{H,H}$ = 7.90 Hz, 2H, 2 x CHar) . ^{13}C-NMR (125 MHz, DMSO-d6): δ [ppm] = 163.0 (1C, CO), 157.6 (1C, Cq), 154.0 (1C, CHole),143.7 (1C, Cq), 132.2 (1C, CHar), 131.2 (2C, 2 x CHar), 130.1 (1C, CHar), 129.2 (1C, Cq), 129.0 (1C, Cq), 128.4 (1C, CHar), 127.1 (1C, CHar), 115.3 (2C,2 x CHar), 106.8 (1C, CN).

(2E)-2-cyano-3-(4'-methyl[1,1'-biphenyl]-4-yl)prop-2-enamide (22)

Pale yellow crystals, 84% yield, Melting Point = 198.2 °C UV-Vis (MeOH: H_2O, 8:2,v/v):λ_{max} = 340 nm MALDI-MS: $[M+H]^+$ = 263.44 Da ^{1}H-NMR (500 MHz, DMSO-d6): δ [ppm] = 8.21 (s, 1H, CHole), 8.03 (d, 3J H,H = 7.90 Hz, 2H, 2 x CHar), 7.92 (s, br, 1H, NH), 7.88 (d, $^3J_{H,H}$ = 8.20 Hz,

2H, 2 x CHar), 7.77 (s, br, 1H, NH), 7.68 (d, $^{3}J_{H,H}$ = 7.90 Hz, 2H, 2 x CHar), 7.32 (d, $^{3}J_{H,H}$ = 7.90 Hz, 2H, 2 x CHar), 2.37 (s, 3H, CH_3). ^{13}C-NMR (125 MHz, DMSO-d6): δ [ppm] = 162.7 (1C, CO), 149.9 (1C, CHole), 143.5 (1C, Cq), 138.0 (1C, Cq), 135.6 (1C, Cq), 130.7 (2C, 2 x CHar), 130.5 (1C, Cq), 129.6 (2C, 2 x CHar), 126.9 (2C, 2 x CHar), 126.6 (2C, 2 x CHar), 116.6 (1C, Cq), 105.7 (1C, CN), 20.6 (1C, CH_3).

(2*E*)-2-cyano-3-(2'-methyl[1,1'-biphenyl]-4-yl)prop-2-enamide (23)

White solid, 84% yield, Melting Point = 145.7 °C UV-Vis (MeOH: H_2O, 8:2,v/v):λ_{max} = 323 nm MALDI-MS: $[M+H]^+$ = 263.47 Da ^{1}H-NMR (500 MHz, DMSO-d6): δ [ppm] = 8.24 (s, 1H, CHole), 8.02 (d, $^{3}J_{H,H}$ = 7.90 Hz, 2H, 2 x CHar), 7.95 (s, br, 1H, NH), 7.80 (s, br, 1H, NH), 7.58 (d, $^{3}J_{H,H}$ = 7.60 Hz, 2H, 2 x CHar), 7.33-7.25 (m, 4H, 4 x CHar), 2.27 (s, 3H, CH_3). ^{13}C-NMR (125 MHz, DMSO-d6): δ [ppm] = 162.6 (1C, CO), 150.1 (1C, CHole), 145.2 (1C, Cq), 139.9 (1C, Cq), 134.7 (1C, Cq), 130.5 (1C, Cq), 130.4 (1C, CHar), 129.9 (2C, 2 x CHar), 129.8 (2C, 2 x CHar), 129.3 (1C, CHar), 127.9 (1C, CHar), 126.0 (1C, CHar), 116.5 (1C, Cq), 106.1 (1C, CN), 20.0 (1C, CH_3).

(2*E*)-2-cyano-3-(4'-methyl[1,1'-biphenyl]-3-yl)prop-2-enamide (24)

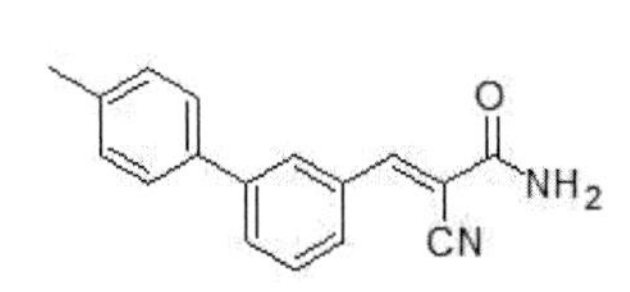

White solid, 84% yield, Melting Point = 154.1 °C

UV-Vis (MeOH: H_2O, 8:2,v/v):λ_{max} = 302 nm MALDI-MS: $[M+H]^+$ = 263.46 Da ^{1}H-NMR (500 MHz, DMSO-d6): δ [ppm] = 8.28 (s, 1H, CHar), 8.22 (s, 1H, CHole), 7.95 (s, 1H, br, NH), 7.90 (d, $^{3}J_{H,H}$ = 7.60 Hz, 1H, CHar), 7.86 (d, $^{3}J_{H,H}$ = 7.60 Hz, 1H, CHar), 7.81 (s, br, 1H, NH), 7.65 (t, $^{3}J_{H,H}$ = 7.90 Hz, 1H, CHar), 7.61 (d, $^{3}J_{H,H}$ = 7.90 Hz, 2H, 2 x CHar), 7.32 (d, $^{3}J_{H,H}$ = 7.60 Hz, 2H, 2 x CHar), 2.36 (s, 3H, CH_3). ^{13}C-NMR

(125 MHz, DMSO-d6): δ [ppm] = 162.6 (1C, CO), 150.6 (1C, CHole), 140.8 (1C, Cq), 137.4 (1C, Cq), 136.1 (1C, Cq), 132.5 (1C, Cq), 130.1 (1C, CHar), 129.7 (1C, CHar), 129.6 (2C, 2 x CHar), 128.4 (1C, CHar), 127.9 (1C, CHar), 126.5 (2C, 2 x CHar), 116.4 (1C, Cq), 106.9 (1C, CN), 20.6 (1C, CH_3).

(2*E*)-3-([1,1'-biphenyl]-4-yl)-2-cyano-N-ethylprop-2-enamide (25)

White crystals, 82% yield, Melting Point = 154.7 °C UV-Vis (MeOH: H_2O, 8:2,v/v):λ_{max} = 327 nm MALDI-MS: $[M+H]^+$ = 277.45 Da ^{1}H-NMR (500 MHz, DMSO-d6): δ [ppm] = 8.50 (s, br, 1H, NH), 8.21 (s, 1H, CHole), 8.05 (d, $^3J_{H,H}$ = 8.20 Hz, 2H, 2 x CHar), 7.90 (d, $^3J_{H,H}$ = 7.90 Hz, 2H, 2 x CHar), 7.78 (d, $^3J_{H,H}$ = 7.50 Hz, 2H, 2 x CHar), 7.52 (t, $^3J_{H,H}$ = 7.40 Hz, 2H, 2 x CHar), 7.45 (t, $^3J_{H,H}$ = 7.05 Hz, 1H, CHar), 3.29-3.23 (m, 2H, CH_2), 1.12 (t, $^3J_{H,H}$ = 7.20 Hz, 3H, CH_3). ^{13}C-NMR (125 MHz, DMSO-d6): δ [ppm] = 160.6 (1C, CO), 149.7 (1C, CHole), 143.5 (1C, Cq), 138.6 (1C, Cq), 130.9 (1C, Cq), 130.7 (2C, 2 x CHar), 129.0 (2C, 2 x CHar), 128.4 (1C, CHar), 127.2 (2C, 2 x CHar), 126.8 (2C, 2 x CHar), 116.5 (1C, Cq), 105.8 (1C, CN), 34.6 (1C, CH_2), 14.4 (1C, CH_3).

(2*E*)-3-([1,1'-biphenyl]-3-yl)-2-cyano-N-ethylprop-2-enamide (26)

White crystals, 77% yield, Melting Point = 124.7 °C UV-Vis (MeOH: H_2O, 8:2,v/v):λ_{max} = 300 nm MALDI-MS: $[M+H]^+$ = 277.37 Da ^{1}H-NMR (500 MHz, DMSO-d6): δ [ppm] = 8.51 (s, 1H, br, NH), 8.27 (s, 1H, CHole), 8.24 (s, 1H, CHar), 7.94 (d, $^3J_{H,H}$ = 8.20 Hz, 1H, CHar), 7.89 (d, $^3J_{H,H}$ = 8.60 Hz, 1H, CHar), 7.72 (d, $^3J_{H,H}$ = 7.60 Hz, 2H, 2 x CHar), 7.67 (t, $^3J_{H,H}$ = 7.40 Hz, 1H, CHar), 7.52 (t, $^3J_{H,H}$ = 7.50 Hz, 2H, 2 x CHar), 7.43 (t, $^3J_{H,H}$ = 7.20 Hz, 1H, CHar), 3.29-3.24 (m, 2H, CH_2), 1.12 (t, $^3J_{H,H}$ = 7.15 Hz, 3H, CH_3) . ^{13}C-NMR (125 MHz, DMSO-d6): δ [ppm] = 160.5 (1C, CO), 150.4 (1C, CHole),140.8 (1C, Cq), 139.0 (1C, Cq), 132.5(1C, Cq), 130.3 (1C, CHar), 129.8 (1C, CHar), 129.0 (2C, 2 x CHar), 128.6 (1C, CHar), 128.2 (1C, CHar), 128.0 (1C, CHar), 126.6 (2C,2 x CHar), 116.4 (1C, Cq), 106.8 (1C, CN), 34.6 (1C, CH_2), 14.4 (1C, CH_3).

(2*E*)-3-(4'-chloro[1,1'-biphenyl]-4-yl)-2-cyanoprop-2-enamide (27)

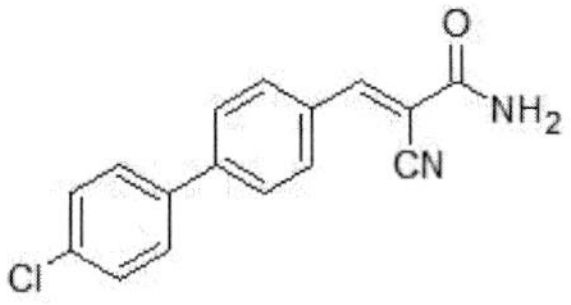

Yellow solid, 83% yield, Melting Point = 286.2 °C UV-Vis (MeOH: H_2O, 8:2,v/v):λ_{max} = 335 nm MALDI-MS: $[M+H]^+$ = 284.17 Da ^{1}H-NMR (500 MHz, DMSO-d6): δ [ppm] = 8.22 (s, 1H, CHole), 8.05 (d, $^3J_{H,H}$ = 8.50 Hz, 2H, 2 x CHar), 7.94 (s, br, 1H, NH), 7.91 (d, $^3J_{H,H}$ = 8.50 Hz, 2H, 2 x CHar), 7.82 (d, $^3J_{H,H}$ = 8.50 Hz, 2H, 2 x CHar), 7.80 (s, br, 1H, NH), 7.57 (d, $^3J_{H,H}$ = 8.5 Hz, 2H, 2 x CHar). ^{13}C-NMR (125 MHz, DMSO-d6): δ [ppm] = 163.2 (1C, CO), 150.4 (1C, CHole), 142.7 (1C, Cq), 138.0 (1C, Cq), 133.9(1C, Cq), 131.7(1C, Cq), 131.3 (2C, 2 x CHar), 129.6 (2C, 2 x CHar), 129.2 (2C, 2 x CHar), 127.8 (2C, 2 x CHar), 117.0 (1C, Cq), 106.9 (1C, CN).

(2*E*)-3-(4'-chloro[1,1'-biphenyl]-3-yl)-2-cyanoprop-2-enamide (28)

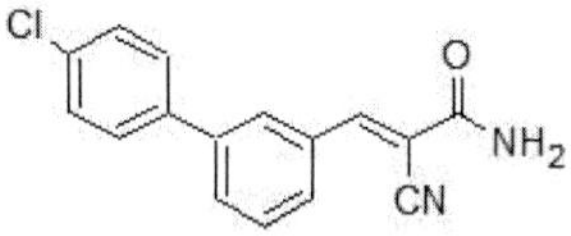

White crystals, 38% yield, Melting Point = 157.4 °C UV-Vis (MeOH: H_2O, 8:2,v/v):λ_{max} = 300 nm MALDI-MS: $[M+H]^+$ = 284.15 Da ^{1}H-NMR (500 MHz, DMSO-d6): δ [ppm] = 8.29 (s, 1H, CHar), 8.22 (s, 1H, CHole), 7.95 (d, $^3J_{H,H}$ = 7.30 Hz, 2H, 2 x CHar), 7.89 (d, $^3J_{H,H}$ = 8.20 Hz, 1H, CHar), 7.81 (s, br, 1H, NH), 7.75 (d, $^3J_{H,H}$ = 8.2 Hz, 2H, 2 x CHar), 7.68 (t, $^3J_{H,H}$ = 7.90 Hz, 1H, CHar), 7.58 (d, $^3J_{H,H}$ = 7.90 Hz, 2H, 2 x CHar). ^{13}C-NMR (125 MHz, DMSO-d6): δ [ppm] = 162.4 (1C, CO), 150.5 (1C, CHole),139.5 (1C, Cq), 137.8 (1C, Cq), 133.0(1C, Cq), 132.5(1C, Cq), 130.2 (1C, CHar), 129.9 (1C, CHar), 128.9 (2C, 2 x CHar), 128.4 (2C, 2 x CHar), 116.4 (1C, Cq), 107.3 (1C, CN).

(2*E*)-2-cyano-3-(4'-methoxy[1,1'-biphenyl]-4-yl)prop-2-enoic acid (29)

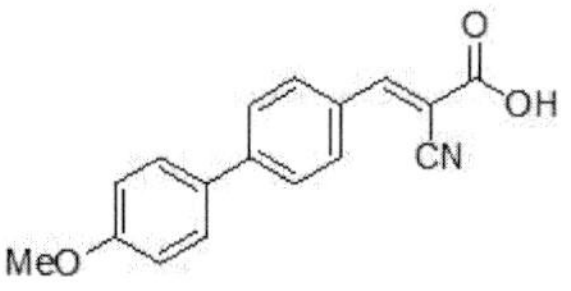

Yellow crystals, 52% yield, Melting Point = 252.6 °C UV-Vis (MeOH: H_2O, 8:2,v/v):λ_{max} = 332 nm MALDI-MS: $[M+H]^+$ = 280.47 Da ^{1}H-NMR (500 MHz, DMSO-d6): δ [ppm] = 8.36 (s, 1H, CHole), 8.12 (d, $^3J_{H,H}$ = 8.20 Hz,

2H, 2 x CHar), 7.88 (d, $^3J_{H,H}$ = 8.50 Hz, 2H, 2 x CHar), 7.77 (d, $^3J_{H,H}$ = 8.20 Hz, 3H, 2 x CHar, NH), 7.07 (d, $^3J_{H,H}$ = 8.80 Hz, 2H, 2 x CHar), 3.82 (s, 3H, CH_3). ^{13}C-NMR (125 MHz, DMSO-d6): δ [ppm] = 163.3 (1C, CO), 159.8 (1C, Cq), 153.7 (1C, CHole), 144.0 (1C, Cq), 131.4 (2C, 2 x CHar), 129.6 (1C, Cq), 128.1 (2C, 2 x CHar), 126.5 (2C, 2 x CHar), 116.3 (1C, Cq), 114.5 (2C, 2 x CHar), 102.5 (1C, CN), 55.2 (1C, CH_3).

(2*E*)-2-cyano-3-(3'-methoxy[1,1'-biphenyl]-4-yl)prop-2-enoic acid (30)

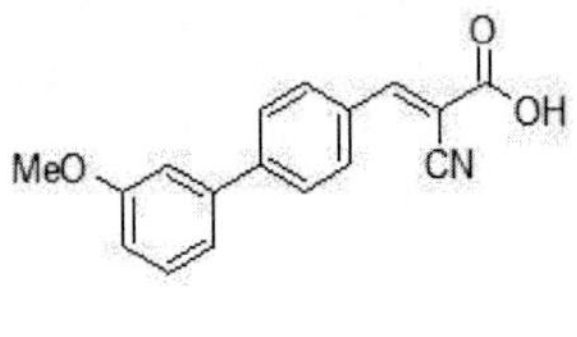

Pale yellow crystals, 91% yield, Melting Point = 198.1 °C UV-Vis (MeOH: H_2O, 8:2,v/v):λ_{max} = 323 nm MALDI-MS: $[M+H]^+$ = 280.31 Da ^{1}H-NMR (500 MHz, DMSO-d6): δ [ppm] = 8.38 (s, 1H, CHole), 8.14 (d, $^3J_{H,H}$ = 8.50 Hz, 2H, 2 x CHar), 7.93 (d, $^3J_{H,H}$ = 7.60 Hz, 2H, 2 x CHar), 7.44-7.41 (m, 1H, CHar), 7.35 (d, $^3J_{H,H}$ = 6.90 Hz, 1H, CHar), 7.31 (s, 1H, CHar), 7.01 (d, $^3J_{H,H}$ = 7.60 Hz, 1H, CHar), 3.84 (s, 3H, CH_3). ^{13}C-NMR (125 MHz, DMSO-d6): δ [ppm] = 163.2 (1C, CO), 159.7 (1C, CHole), 153.6 (1C, CHole), 144.2 (1C, Cq), 139.9 (1C, Cq), 131.2 (2C, 2 x CHar), 130.6 (1C, Cq), 130.1 (1C, CHar), 127.4 (2C, 2 x CHar), 119.2 (1C, CHar), 116.2 (1C, Cq), 114.2 (1C, CHar), 112.3 (1C, CHar), 103.3 (1C, CN), 55.1 (1C, CH_3).

(2*E*)-2-cyano-3-(2'-methoxy[1,1'-biphenyl]-4-yl)prop-2-enoic acid (31)

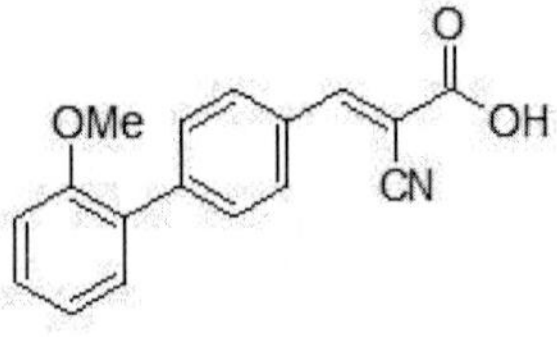

Yellow crystals, 86% yield, Melting Point = 197.8 °C UV-Vis (MeOH: H_2O, 8:2,v/v):λ_{max} = 328 nm MALDI-MS: $[M+H]^+$ = 280.29 Da ^{1}H-NMR (500 MHz, DMSO-d6): δ [ppm] = 8.35 (s, 1H, CHole), 8.08 (d, $^3J_{H,H}$ = 8.20 Hz, 2H, 2 x CHar), 7.70 (d, $^3J_{H,H}$ = 7.90 Hz, 2H, 2 x CHar), 7.42-7.37 (m, 2H, 2 x CHar), 7.16 (d, $^3J_{H,H}$ = 7.60 Hz, 1H, CHar), 7.07 (t, $^3J_{H,H}$ = 7.30 Hz, 1H, CHar), 3.80 (s, 3H, CH_3). ^{13}C-NMR (125 MHz, DMSO-d6): δ [ppm] = 163.3 (1C, CO), 156.2 (1C, Cq), 153.8 (1C, CHole), 130.3 (2C, 2 x CHar), 130.2 (1C, CHar), 129.9 (2C, 2 x CHar), 129.8 (1C, CHar), 129.0 (1C, CHar), 128.2 (1C, Ci), 120.8 (1C, CHar), 111.8 (1C, CHar), 103.8 (1C, CN), 55.5 (1C, CH_3).

(2*E*)-3-(4'-chloro-3'-fluoro[1,1'-biphenyl]-4-yl)-2-cyanoprop-2-enamide (32)

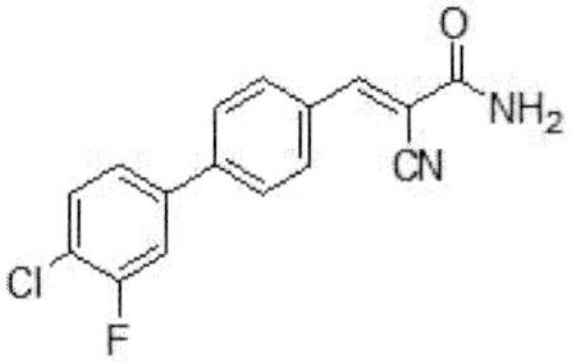

Pale yellow crystals, 73% yield, Melting Point = 217.7 °C UV-Vis (MeOH: H_2O, 8:2,v/v):λ_{max} = 325 nm MALDI-MS: $[M+H]^+$ = 302.05 Da ^{1}H-NMR (500 MHz, DMSO-d6): δ [ppm] = 8.40 (s, 1H, CHole), 8.15 (d, $^3J_{H,H}$ = 7.90 Hz, 2H, 2 x CHar), 7.96 (d, $^3J_{H,H}$ = 7.30 Hz, 3H, 2 x CHar, NH), 7.91 (d, $^3J_{H,F}$ = 10.70 Hz, 1H, CHar), 7.81 (s, br, 1H, NH), 7.73-7.68 (m, 2H, 2 x CHar). ^{13}C-NMR (125 MHz, DMSO-d6): Ît' [ppm] = 163.1 (1C, CO), 158.5 (1C, Cq), 156.6 (1C, Cq), 153.3 (1C, CHole), 140.9 (d, $^1J_{C,F}$ = 247 Hz, 1C, CF), 131.3 (1C, Cq), 131.2 (2C, 2 x CHar), 131.1 (1C, CHar), 127.4 (2C, 2 x CHar), 124.0 (1C, CHar), 116.1 (d, $^1J_{C,F}$ = 24 Hz, 1C, Cq), 115.3 (d, $^2J_{C,F}$ = 21 Hz, 1C, CHar), 104.3 (1C, CN).

(2*E*)-3-([1,1'-biphenyl]-2-yl)-2-cyanoprop-2-enoic acid (34)

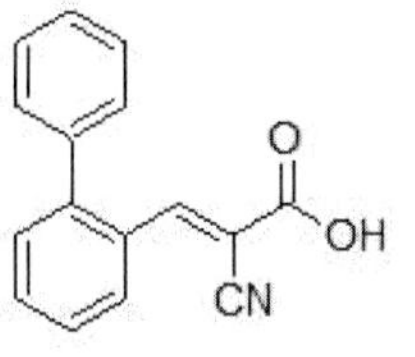

White crystals, 76% yield, Melt-ing Point = 184.4 °C UV-Vis (MeOH: H_2O, 8:2,v/v):λ_{max} = 295 nm MALDI-MS: $[M+H]^+$ = 250.37 Da ^{1}H-NMR (500 MHz, DMSO-d6): δ [ppm] = 8.15 (d, $^3J_{H,H}$ = 7.60 Hz, 1H, CHar), 8.05 (s, 1H, CHole), 7.71-7.68 (m, 1H, CHar), 7.64-7.61 (m, 1H, CHar), 7.57 (d, $^3J_{H,H}$ = 6.90 Hz, 1H, CHar), 7.53-7.47 (m, 3H, 3 x CHar), 7.36 (d, $^3J_{H,H}$ = 6.90 Hz, 2H, 2 x CHar). ^{13}C-NMR (125 MHz, DMSO-d6): δ [ppm] = 163.4 (1C, CO), 154.3 (1C, CHole), 144.1 (1C, Cq), 139.1 (1C, Cq), 132.9 (1C, CHar), 130.9 (1C, CHar), 130.3 (1C, CHar), 130.0 (1C, Cq), 129.1 (1C, CHar), 129.0 (1C, CHar), 128.7 (1C, CHar), 128.5 (1C, CHar), 116.3 (1C, Cq), 106.2 (1C, CN).

(2*E*)-2-cyano-3-[4'-(trifluoromethyl)[1,1'-biphenyl]-4-yl]prop-2-enamide (35)

White crystals, 85% yield, Melting Point = 203.7 °C UV-Vis (MeOH: H_2O, 8:2,v/v):λ_{max} = 326 nm MALDI-MS: $[M+H]^+$ = 317.56 Da ^{1}H-NMR (500 MHz, DMSO-d6): δ [ppm] = 8.25 (s, 1H, CHole), 8.09 (d, $^3J_{H,H}$ = 7.90 Hz, 2H, 2 x CHar), 8.01 (d, $^3J_{H,H}$ = 8.20 Hz, 2H, 2 x CHar), 7.98 (d, $^3J_{H,H}$ = 7.90 Hz, 3H, 2 x CHar, NH), 7.87 (d, $^3J_{H,H}$ = 8.20 Hz, 2H, 2 x CHar), 7.83 (s, br, 1H, NH). ^{13}C-NMR (125 MHz, DMSO-d6): δ [ppm] = 162.6 (1C, CO), 149.7 (1C, CHole), 142.6 (1C, Cq), 141.8 (1C, Cq), 131.8 (1C, Cq), 130.7(2C, 2 x CHar), 128.4 (1C, Cq), 127.7 (2C, 2 x CHar), 125.8 (1C, CF_3), 124.1 (q, , $^1J_{C,F}$ = 271 Hz, 1C, CF_3), 123.1(1C, Cq), 116.4 (1C, Cq), 106.8 (1C, CN).

(2E)-2-cyano-3-[2'-(trifluoromethyl)[1,1'-biphenyl]-4-yl]prop-2-enamide (36)

White crystals, 93% yield, Melting Point = 138.6 °C UV-Vis (MeOH: H_2O, 8:2,v/v):λ_{max} = 312 nm MALDI-MS: $[M+H]^+$ = 317.39 Da ^{1}H-NMR (500 MHz, DMSO-d6): δ [ppm] = 8.25 (s, 1H, CHole), 8.01 (d, $^3J_{H,H}$ = 7.90 Hz, 2H, 2 x CHar), 7.97 (s, br, 1H, NH), 7.88 (d, $^3J_{H,H}$ = 7.90 Hz, 1H, CHar), 7.82 (s, br, 1H, NH), 7.77-7.75 (m, 1H, CHar), 7.69-7.65 (m, 1H, CHar), 7.53 (d, $^3J_{H,H}$ = 7.60 Hz, 2H, 2 x CHar), 7.47 (d, $^3J_{H,H}$ = 7.60 Hz, 1H, CHar). ^{13}C-NMR (125 MHz, DMSO-d6): δ [ppm] = 161.7 (1C, CO), 153.5 (1C, CHole), 144.7 (1C, Cq), 139.8 (1C, Cq), 131.6 (1C, CHar), 130.9 (1C, Cq), 130.3 (4C, 4 x CHar), 130.0 (1C, CHar), 128.1 (1C, CHar), 126.3 (1C, Cq), 123.9 (q, $^1J_{C,F}$ = 279 Hz, 1C, CF_3), 117.1 (1C, Cq), 103.2 (1C, CN).

(2*E*)-2-cyano-3-[4'-(trifluoromethyl)[1,1'-biphenyl]-3-yl]prop-2-enamide (37)

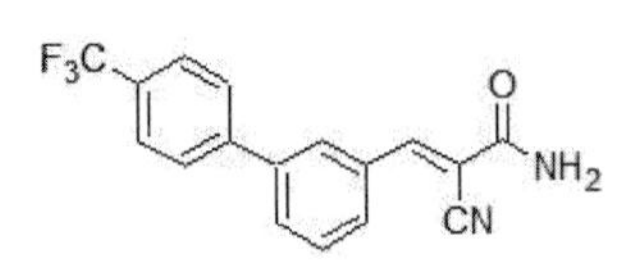

White crystals, 67% yield, Melting Point = 142.9 °C UV-Vis (MeOH: H_2O, 8:2,v/v):λ_{max} = 297 nm MALDI-MS: $[M+H]^+$ = 317.40 Da ^{1}H-NMR (500 MHz, DMSO-d6): δ [ppm] = 8.31 (s, 1H, CHar), 8.29 (s, 1H, CHole), 8.01 (d, $^3J_{H,H}$ = 7.60 Hz, 1H, CHar), 7.97-

7.94 (m,4H, 3 x CHar, NH), 7.89-7.88 (m, 2H, 2 x CHar), 7.83 (s, br, 1H, NH), 7.72 (t, 3J H,H = 7.90 Hz, 1H, CHar). ^{13}C-NMR (125 MHz, DMSO-d6): δ [ppm] = 162.4 (1C, CO), 150.3 (1C, CHole), 142.9 (1C, Cq), 139.3 (1C, Cq), 132.7 (1C, Cq), 130.6 (1C, CHar), 130.0 (1C, CHar), 129.4 (1C, CHar), 128.7 (1C, CHar), 127.5 (2C, 2 x CHar), 125.9 (1C, Cq), 122.7 (q, $^{1}J_{C,F}$ = 271 Hz, 1C, CF_3), 116.3 (1C, Cq), 107.4 (1C, CN).

(2*E*)-2-cyano-3-[2'-(trifluoromethyl)[1,1'-biphenyl]-3-yl]prop-2-enamide (38)

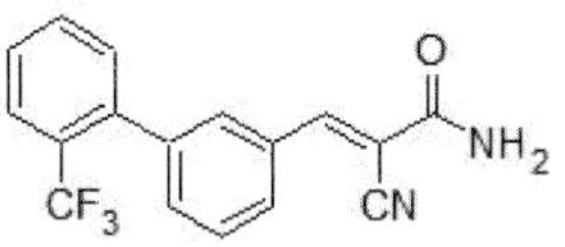

CF3 CN O NH2

White crystals, 85% yield, Melting Point = 117.2 °C UV-Vis (MeOH: H_2O, 8:2,v/v):λ_{max} = 298 nm MALDI-MS: $[M+H]^+$ = 317.25 Da ^{1}H-NMR (500 MHz, DMSO-d6): δ [ppm] = 8.23 (s, 1H, CHole), 8.01 (d, $^{3}J_{H,H}$ = 6.90 Hz, 1H, CHar), 7.95 (s, br, 1H, NH), 7.89-7.87 (m, 2H, 2 x CHar), 7.80 (s, br, 1H, NH), 7.77-7.74 (m, 1H, CHar), 7.69-7.64 (m, 2H, 2 x CHar), 7.58 (s, 1H, CHar), 7.46 (t, $^{3}J_{H,H}$ = 6.80 Hz, 1H, CHar). ^{13}C-NMR (125 MHz, DMSO-d6): δ [ppm] = 163.1 (1C, CO), 150.5 (1C, CHole), 139.9 (1C, Cq), 133.4 (1C, CHar), 133.0 (1C, CHar), 132.9 (1C, CHar), 132.6 (1C, CHar), 132.5 (1C, CHar), 130.4 (1C, CHar), 130.0(1C, CHar), 129.8 (1C, CHar), 129.4 (1C, CHar), 129.2 (1C, CHar), 128.9 (1C, Cq), 126.6 (1C, Cq), 122.8 (q, $^{1}J_{C,F}$ = 271 Hz, 1C, CF_3), 107.8 (1C, CN).

(2*E*)-2-cyano-3-[4'-(trifluoromethyl)[1,1'-biphenyl]-2-yl]prop-2-enamide (39)

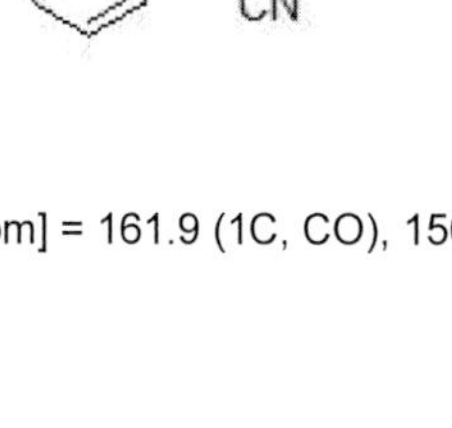

CF3 O NH2 CN

Yellow solid, 75% yield, Melting Point = 168.0 °C UV-Vis (MeOH: H_2O, 8:2,v/v):λ_{max} = 302 nm MALDI-MS: $[M+H]^+$ = 317.42 Da ^{1}H-NMR (500 MHz, DMSO-d6): δ [ppm] = 8.04 (d, $^{3}J_{H,H}$ = 7.90 Hz, 1H, CHar), 7.97 (s, 1H, CHole), 7.87 (d, $^{3}J_{H,H}$ = 7.90 Hz, 2H, 2 x CHar), 7.84 (s, 1H, br, NH), 7.74 (s, br, 1H, NH), 7.70-7.63 (m, 2H, 2 x CHar), 7.60-7.57 (m, 3H, 3 x CHar) . ^{13}C-NMR (125 MHz, DMSO-d6): δ [ppm] = 161.9 (1C, CO), 150.2 (1C, CHole), 142.9 (1C, Cq), 141.1 (1C, Cq),

131.7 (1C, CHar), 130.5 (2C, 2 x CHar), 130.3 (1C, CHar), 130.2 (1C, Cq), 128.7 (1C, CHar), 128.5 (1C, CHar), 125.3 (1C, Cq), 123.1 (q, $^1J_{C,F}$ = 269 Hz, 1C, CF_3), 116.0 (1C, Cq), 109.5 (1C, CN).

(2*E*)-2-cyano-3-(9*H*-fluoren-2-yl)prop-2-enoic acid (40)

Yellow crystals, 82% yield, Melting Point = 260.2 °C UV-Vis (MeOH: H_2O, 8:2,v/v):λ_{max} = 342 nm MALDI-MS: $[M+H]^+$ = 262.4 Da ^{1}H-NMR (500 MHz, DMSO-d6): δ [ppm] = 13.89 (s, br, 1H, COOH), 8.40 (s, 1H, CHar), 8.31 (s, 1H, CHole), 8.13-8.09 (m, 2H, 2 x CHar), 8.04 (d, $^3J_{H,H}$ = 7.30 Hz, 1H, CHar), 7.66 (d, $^3J_{H,H}$ = 7.30 Hz, 1H, CHar), 7.47-7.42 (m, 2H, 2 x CHar), 4.04 (s, 2H, CH_2). ^{13}C-NMR (125 MHz, DMSO-d6): δ [ppm] = 163.5 (1C, CO), 154.5 (1C, CHole), 146.0 (1C, Cq), 144.4 (1C, Cq), 143.7 (1C, Cq), 139.7 (1C,Cq), 130.5 (1C, CHar), 129.8 (1C, Cq), 128.4 (1C, Cq), 127.0 (1C, CHar), 126.9 (1C, CHar), 125.3 (1C, CHar), 121.2 (1C, CHar), 120.6 (1C, CHar), 116.4 (1C, Cq), 101.9 (1C, CN), 36.3 (1C, CH_2).

(2*E*)-2-cyano-3-(4'-fluoro[1,1'-biphenyl]-4-yl)prop-2-enoic acid (41)

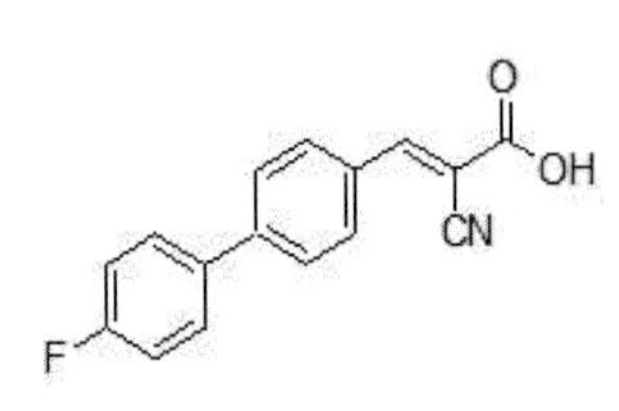

Yellow crystals, 52% yield, Melting Point = 243.3 °C UV-Vis (MeOH: H_2O, 8:2,v/v):λ_{max} = 323 nm MALDI-MS: $[M+H]^+$ = 267.58 Da ^{1}H-NMR (500 MHz, DMSO-d6): δ [ppm] = 14.10 (s, br, 1H, COOH), 8.38 (s, 1H, CHole), 8.14 (d, $^3J_{H,H}$ = 8.25 Hz, 2H, 2 x CHar), 7.91 (d, $^3J_{H,H}$ = 8.35 Hz, 2H, 2 x CHar), 7.87-7.84 (m, 2H, 2 x CHar), 7.35 (t, $^3J_{H,H}$ = 8.60 Hz, 2H, 2 x CHar). ^{13}C-NMR (125 MHz, DMSO-d6): δ [ppm] = 163.2 (1C, CO), 162.5 (d, $^1J_{C,F}$ = 240 Hz, 1C, CF), 153.5 (1C, CHole), 143.1 (1C, Cq), 134.9 (1C, Cq), 131.3 (2C, 2 x CHar), 130.4 (1C, Cq), 129.1 (2C, 2 x CHar), 129.0 (1C, Cq), 127.2 (2C, 2 x CHar), 116.2 (1C, Cq), 115.8 (d, $^2J_{C,F}$ = 21 Hz, 2C, 2 x CHar), 103.3 (1C, CN).

(2*E*)-2-cyano-3-(3'-fluoro[1,1'-biphenyl]-3-yl)prop-2-enoic acid (42)

White crystals, 51% yield, Melting Point = 188.1 °C UV-Vis (MeOH: H_2O, 8:2,v/v):λ_{max} = 285 nm MALDI-MS: $[M+H]^+$ = 267.48 Da ^{1}H-NMR (500 MHz, DMSO-d6): δ [ppm] = 8.46 (s, 1H, CHole), 8.36 (s, 1H, CHar), 8.08 (d, $^3J_{H,H}$ = 7.90 Hz, 1H, CHar), 7.98 (d, $^3J_{H,H}$ = 8.50 Hz, 1H, CHar), 7.70 (t, $^3J_{H,H}$ = 8.0 Hz, 1H, CHar), 7.61-7.55 (m, 3H, 3 x CHar), 7.28-7.25 (m, 1H, CHar). ^{13}C-NMR (125 MHz, DMSO-d6): δ [ppm] = 163.0 (1C, CO), 162.6 (d, $^1J_{C,F}$ = 238 Hz, 1C, CF), 154.2 (1C, CHole), 141.2 (1C, Cq), 139.4 (1C, Cq), 132.2 (1C, Cq), 131.1 (1C, CHar), 131.0 (1C, CHar), 129.9 (1C, CHar), 129.4 (1C, CHar), 129.2 (1C, CHar), 122.7 (1C, CHar), 116.1 (1C, Cq), 114.7 (d, 2J C,F = 22 Hz, 1C, CHar), 113.3 ((d, $^2J_{C,F}$ = 22 Hz, 1C, CHar), 104.6 (1C, CN).

(2*E*)-2-cyano-N,N-dimethyl-3-(4'-methyl[1,1'-biphenyl]-4-yl)prop-2-enamide (43)

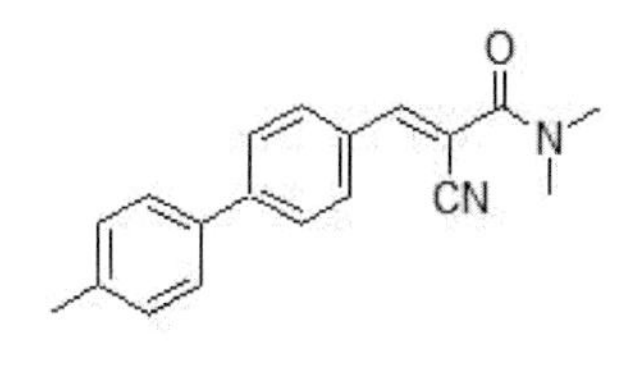

White crystals, 84% yield, Melting Point = 135.2 °C UV-Vis (MeOH: H_2O, 8:2,v/v):λ_{max} = 333 nm MALDI-MS: $[M+H]^+$ = 291.28 Da

^{1}H-NMR (500 MHz, CDCl3): δ [ppm] = 7.61 (d, $^3J_{H,H}$ = 7.60 Hz, 2H, 2 x CHar), 7.50 (d, $^3J_{H,H}$ = 8.20 Hz, 2H, 2 x CHar), 7.47 (d, $^3J_{H,H}$ = 8.50 Hz, 2H, 2 x CHar), 7.34 (s, 1H, CHole), 7.27 (d, $^3J_{H,H}$ = 8.25 Hz, 2H, 2 x CHar), 3.09 (s, 3H, CH_3), 2.99 (s, 3H, CH_3), 2.40 (s, 3H, CH_3).^{13}C-NMR (125 MHz, $CDCl_3$): δ [ppm] = 162.6 (1C, CO), 146.3 (1C, CHole), 138.3 (1C, Cq), 136.6 (1C, Cq), 132.3 (1C, Cq), 131.0 (1C, C),

129.8 (2C, 2 x CHar), 129.7 (2C, 2 x CHar), 127.5 (2C, 2 x CHar), 126.9 (2C, 2 x CHar), 116.8 (1C, Cq), 108.1 (1C, CN), 37.8 (1C, CH_3), 35.1 (1C, CH_3), 21.2 (1C, CH_3).

(2*E*)-2-cyano-3-(2',4'-difluoro[1,1'-biphenyl]-4-yl)prop-2-enoic acid (44)

Pale yellow crystals, 27% yield, Melting Point = 244.7 °C UV-Vis (MeOH: H_2O, 8:2,v/v):λ_{max} = 315 nm

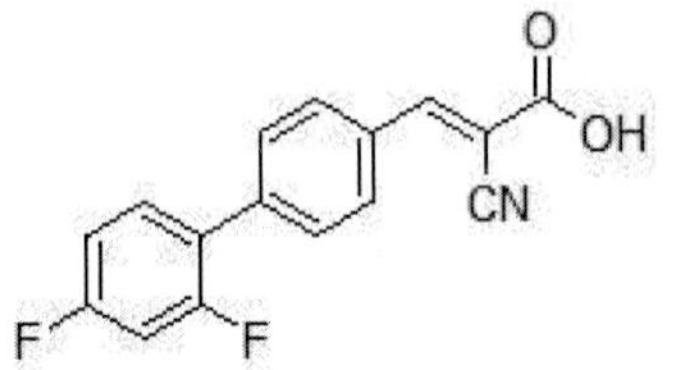

MALDI-MS: $[M+H]^+$ = 286.31 Da
^{1}H-NMR (500 MHz, DMSO-d6): δ [ppm] = 8.06 (s, 1H, CHole), 8.02 (d, $^3J_{H,H}$ = 8.20 Hz, 2H, 2 x CHar), 7.68 (d, $^3J_{H,H}$ = 7.90 Hz, 2H, 2 x CHar), 7.40 (ddd, $^3J_{H,F}$ = 11.3 Hz, $^3J_{H,H}$ = 9.20 Hz, $^4J_{H,F}$ = 2.5 Hz 1H, CHar), 7.24 (td, $^3J_{H,F}$ = 8.60 Hz, $^4J_{H,F}$ = 2.70 Hz, 1H, CHar). ^{13}C-NMR (125 MHz, DMSO-d6): δ [ppm] = 163.7 (1C, CO), 162.7 (dd, $^1J_{C,F}$ = 249 Hz, $^3J_{C,F}$ = 12 Hz, 1C, CF), 159.7 ((dd, $^1J_{C,F}$ = 251 Hz, $^3J_{C,F}$ = 12 Hz, 1C, CF), 154.4 (1C, CHole), 138.4 (1C, Cq), 132.4 (dd, $^3J_{C,F}$ = 10 Hz, $^4J_{C,F}$ = 4.5 Hz, 1C, CHar), 131.8 (1C, Cq), 130.9 (2C, 2 x CHar), 129.8 (2C, 2 x CHar), 124.2 (1C, Cq), 116.4 (1C, Cq), 112.7 (d, , $^1J_{C,F}$ = 18.3 Hz, 1C, CHar), 106.6 (1C, CN), 105.2 (t,$^2J_{C,F}$ = 25 Hz, 1C, CHar).

(2*E*)-2-cyano-*N*-ethyl-3-(4'-methyl[1,1'-biphenyl]-4-yl)prop-2-enamide (45)

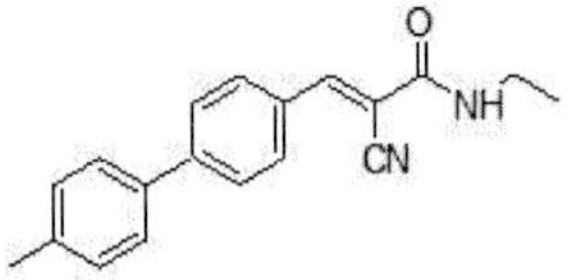

White crystals, 84% yield, Melting Point = 165.2 °C UV-Vis (MeOH: H_2O, 8:2,v/v):λ_{max} = 340 nm MALDI-MS: $[M+H]^+$ = 291.42 Da ^{1}H-NMR (500 MHz, DMSO-d6): δ [ppm] = 8.47 (s, br, 1H, NH), 8.19 (s, 1H, CHole), 8.03 (d, $^3J_{H,H}$ = 7.90 Hz, 2H, 2 x CHar), 7.87 (d, $^3J_{H,H}$ = 7.90 Hz, 2H, 2 x CHar), 7.68 (d, $^3J_{H,H}$ = 7.90 Hz, 2H, 2 x CHar), 7.32 (d, $^3J_{H,H}$ = 7.60 Hz, 2H, 2 x CHar), 3.28-3.23 (m, 2H, CH_2), 2.36 (s, 3H, CH_3), 1.12 (t, $^3J_{H,H}$ = 7.10 Hz, 3H, CH_3). ^{13}C-NMR (125 MHz, DMSO-d6): δ [ppm] = 160.6 (1C, CO), 149.7 (1C, CHole), 143.5 (1C, Cq), 137.9 (1C, Cq), 135.7 (1C, Cq), 130.7 (2C, 2 x CHar), 130.5 (1C, Cq), 129.6 (2C, 2 x CHar), 126.9 (2C, 2 x CHar), 126.6 (2C, 2 x CHar), 116.5 (1C, Cq), 105.6 (1C, CN), 34.6 (1C, CH_2), 20.6 (1C, CH_3), 14.4 (1C, CH_3).

(2*E*)-2-cyano-*N*-ethyl-3-(4'-methyl[1,1'-biphenyl]-3-yl)prop-2-enamide (46)

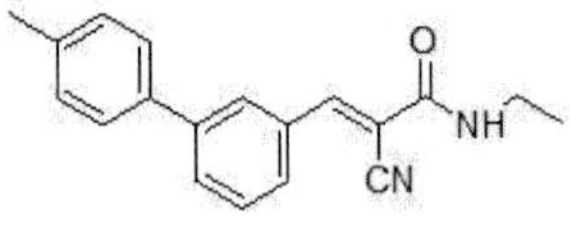

White crystals, 77% yield, Melting Point = 153.5 °C UV-Vis (MeOH: H_2O, 8:2,v/v):λ_{max} = 301 nm MALDI-MS: $[M+H]^+$ = 291.53 Da
^{1}H-NMR (500 MHz, DMSO-d6): δ [ppm] = 8.50 (s, 1H, br, NH), 8.25 (s, 1H, CHole), 8.22 (s, 1H, CHar), 7.90

(d, $^3J_{H,H}$ = 7.20 Hz, 1H, CHar), 7.85 (d, $^3J_{H,H}$ = 7.95 Hz, 1H, CHar), 7.64 (t, $^3J_{H,H}$ = 7.65 Hz, 1H, CHar), 7.61 (d, $^3J_{H,H}$ = 8.00 Hz, 2H, 2 x CHar), 7.32 (d, $^3J_{H,H}$ = 7.85 Hz, 2H, 2 x CHar), 3.26 (q, $^3J_{H,H}$ = 6.37 Hz, 2H, CH_2), 2.36 (s, 3H, CH_3), 1.12 (t, $^3J_{H,H}$ = 7.12 Hz, 3H, CH_3) . ^{13}C-NMR (125 MHz, DMSO-d6): δ [ppm] = 160.5 (1C, CO), 150.4 (1C, CHole), 140.8 (1C, Cq), 137.4 (1C, Cq), 136.1 (1C, Cq), 132.5 (1C, Cq), 130.0 (1C, CHar), 129.7 (1C, CHar), 129.6 (2C, 2 x CHar), 128.3 (1C, CHar), 127.9 (1C, CHar), 126.5 (2C, 2 x CHar), 116.4 (1C, Cq), 106.7 (1C, CN), 34.6 (1C, CH_3), 20.6 (1C, CH_3), 14.4 (1C, CH_3).

(2*E*)-2-cyano-3-(4'-methyl[1,1'-biphenyl]-4-yl)prop-2-enoic acid (47)

Yellow crystals, 81% yield, Melting Point = 241.9 °C UV-Vis (MeOH: H_2O, 8:2,v/v):λ_{max} = 331 nm MALDI-MS: $[M+H]^+$ = 264.33 Da ^{1}H-NMR (500 MHz, DMSO-d6): δ [ppm] = 14.01 (s, br, 1H, COOH), 8.37 (s, 1H, CHole), 8.13 (d, $^3J_{H,H}$ = 8.20 Hz, 2H, 2 x CHar), 7.90 (d, $^3J_{H,H}$ = 8.20 Hz, 2H, 2 x CHar), 7.70 (d, $^3J_{H,H}$ = 7.90 Hz, 2H, 2 x CHar), 7.32 (d, $^3J_{H,H}$ = 7.90 Hz, 2H, 2 x CHar), 2.37 (s, 3H, CH_3). ^{13}C-NMR (125 MHz, DMSO-d6): δ [ppm] = 163.3 (1C, CO), 153.7 (1C, CHole), 144.3 (1C, Cq), 138.2 (1C, Cq), 135.5 (1C, Cq), 131.3 (2C, 2 x CHar), 130.1 (1C, Cq), 129.6 (2C, 2 x CHar), 126.9 (2C, 2 x CHar), 126.7 (2C, 2 x CHar), 116.2 (1C, Cq), 102.8 (1C, CN), 20.6 (1C, CH_3).

(2*E*)-2-cyano-3-(2'-methyl[1,1'-biphenyl]-4-yl)prop-2-enoic acid (48)

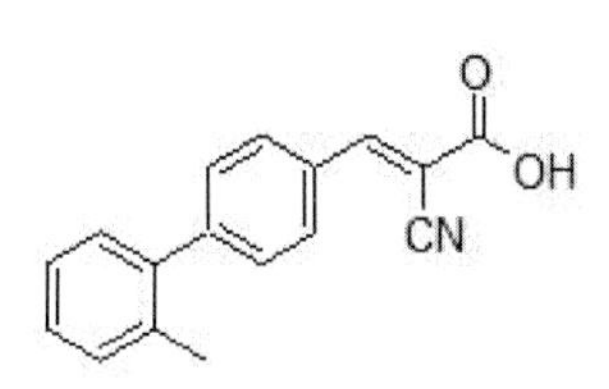

Pale yellow solid, 84% yield, Melting Point = 139.3 °C UV-Vis (MeOH: H_2O, 8:2,v/v):λ_{max} = 309 nm MALDI-MS: $[M+H]^+$ = 264.41 Da

^{1}H-NMR (500 MHz, DMSO-d6): δ [ppm] = 8.39 (s, 1H, CHole), 8.12 (d, $^3J_{H,H}$ = 7.90 Hz, 2H, 2 x CHar), 7.58 (d, $^3J_{H,H}$ = 7.60 Hz, 2H, 2 x CHar), 7.33-7.25 (m, 4H, 4 x CHar), 2.27 (s, 3H, CH_3). ^{13}C-NMR (125 MHz, DMSO-d6): δ [ppm] = 163.2 (1C, CO), 153.7 (1C, CHole), 145.9 (1C, Cq), 139.9 (1C, Cq), 134.7 (1C, Cq), 130.6 (2C, 2 x CHar), 130.5 (1C, CHar), 130.1 (1C, Cq), 129.8 (2C, 2 x CHar), 129.3 (1C, CHar), 127.9 (1C, CHar), 126.0 (1C, CHar), 116.2 (1C, Cq), 20.0 (1C, CH_3).

(2*E*)-2-cyano-3-(4'-methyl[1,1'-biphenyl]-3-yl)prop-2-enoic acid (49)

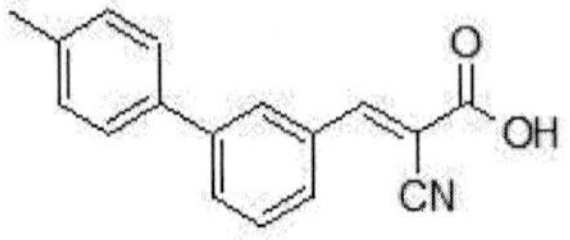

Pale yellow solid, 77% yield, Melting Point = 176.7 °C UV-Vis (MeOH: H_2O, 8:2,v/v):λ_{max} = 263 nm MALDI-MS: $[M+H]^+$ = 264.47 Da ^{1}H-NMR (500 MHz, DMSO-d6): δ [ppm] = 8.44 (s, 1H, CHar), 8.33 (s, 1H, CHole), 8.01 (d, $^3J_{H,H}$ = 7.60 Hz, 1H, CHar), 7.91 (d, $^3J_{H,H}$ = 6.60 Hz, 1H, CHar), 7.68-7.65 (m, 1H, CHar), 7.63 (d, $^3J_{H,H}$ = 7.60 Hz, 2H, 2 x CHar), 7.33 (d, $^3J_{H,H}$ = 7.60 Hz, 2H, 2 x CHar), 2.37 (s, 3H, CH_3). ^{13}C-NMR (125 MHz, DMSO-d6): δ [ppm] = 163.1 (1C, CO), 154.1 (1C, CHole), 140.8 (1C, Cq), 137.4 (1C, Cq), 135.9 (1C, Cq), 132.1 (1C, Cq), 130.7 (1C, CHar), 129.8 (1C, CHar), 129.6 (2C, 2 x CHar), 128.8 (1C, CHar), 128.5 (1C, CHar), 126.5 (2C, 2 x CHar), 116.2 (1C, Cq), 101.8 (1C, CN), 20.6 (1C, CH_3).

(2*E*)-3-(4'-chloro[1,1'-biphenyl]-4-yl)-2-cyanoprop-2-enoic acid (50)

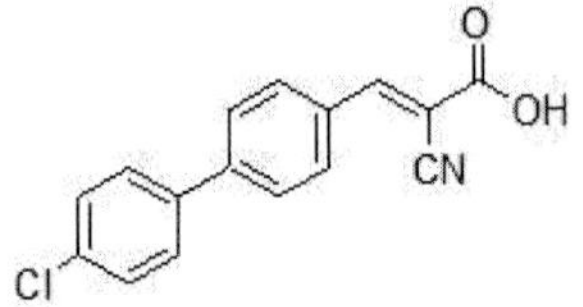

Yellow crystals, 68% yield, Melting Point = 264.9 °C UV-Vis (MeOH: H_2O, 8:2,v/v):λ_{max} = 324 nm MALDI-MS: $[M+H]^+$ = 285.12 Da ^{1}H-NMR (500 MHz, DMSO-d6): δ [ppm] = 13.99 (s, br, 1H, COOH), 8.39 (s, 1H, CHole), 8.15 (d, $^3J_{H,H}$ = 8.50 Hz, 2H, 2 x CHar), 7.93 (d, $^3J_{H,H}$ = 8.20 Hz, 2H, 2 x CHar), 7.83 (d, $^3J_{H,H}$ = 8.20 Hz, 2H, 2 x CHar), 7.58 (d,$^3J_{H,H}$ = 8.50 Hz, 2H, 2 x CHar). ^{13}C-NMR (125 MHz, DMSO-d6): δ [ppm] = 163.2 (1C, CO), 153.5 (1C, CHole), 142.8 (1C, Cq), 137.3 (1C, Cq), 133.4 (1C, Cq), 131.3 (2C, 2 x CHar), 130.8 (1C, Cq), 129.0 (2C, 2 x CHar), 128.7 (2C, 2 x CHar), 127.2 (2C, 2 x CHar), 116.1 (1C, Cq), 103.5 (1C, CN).

(2*E*)-3-(4'-chloro[1,1'-biphenyl]-3-yl)-2-cyanoprop-2-enoic acid (51)

Pale yellow crystals, 79% yield, Melting Point = 204.4 °C UV-Vis (MeOH: H_2O, 8:2,v/v):λ_{max} = 262 nm

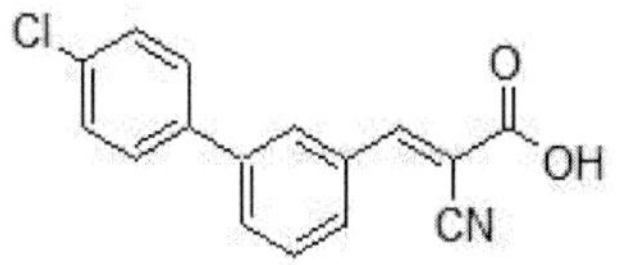

MALDI-MS: $[M+H]^+$ = 286.13 Da ^{1}H-NMR (500 MHz, DMSO-d6): δ [ppm] = 14.11 (s, br, 1H, COOH), 8.45 (s, 1H, CHar), 8.34 (s, 1H, CHole), 8.07 (d, $^3J_{H,H}$ = 7.60 Hz, 1H, CHar), 7.94 (d, $^3J_{H,H}$ = 7.60 Hz, 1H, CHar), 7.76 (d, $^3J_{H,H}$ = 8.20 Hz, 2H, 2 x CHar), 7.69 (t, $^3J_{H,H}$ = 7.90 Hz, 1H, CHar), 7.58 (d, $^3J_{H,H}$ = 8.20 Hz, 2H, 2 x CHar). ^{13}C-NMR (125 MHz, DMSO-d6): δ [ppm] = 163.6 (1C, CO), 154.7 (1C, CHole), 140.1 (1C, Cq), 138.2 (1C, Cq), 133.5 (1C, Cq), 132.8 (1C, Cq), 131.5 (1C, CHar), 130.5 (1C, CHar), 129.8 (1C, Cq), 129.6 (2C, 2 x CHar), 129.0 (2C, 2 x CHar), 116.6 (1C, Cq), 105.1 (1C, CN).

(2*E*)-3-(4'-chloro-3'-fluoro[1,1'-biphenyl]-4-yl)-2-cyanoprop-2-enoic acid (52)

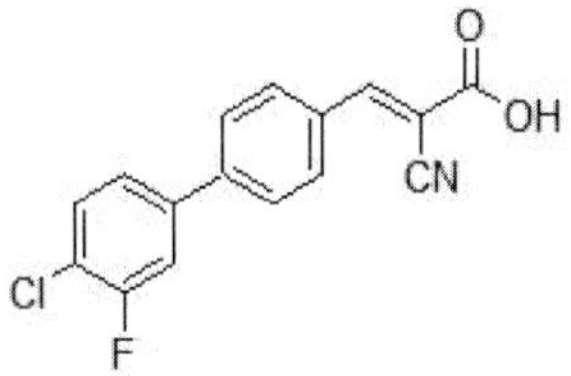

Yellow crystals, 61% yield, Melting Point = 259.8 °C UV-Vis (MeOH: H_2O, 8:2,v/v):λ_{max} = 329 nm MALDI-MS: $[M+H]^+$ = 303.07 Da ^{1}H-NMR (500 MHz, DMSO-d6): δ [ppm] = 8.39 (s, 1H, CHole), 8.15 (d, $^3J_{H,H}$ = 7.90 Hz, 2H, 2 x CHar), 7.98 (d, $^3J_{H,H}$ = 7.90 Hz, 2H, 2 x CHar), 7.92 (d, $^3J_{H,F}$ = 9.10 Hz, 1H, CHar), 7.73-7.69 (m, 2H, 2 x CHar). ^{13}C-NMR (125 MHz, DMSO-d6): δ [ppm] = 163.7 (1C, CO), 153.8 (1C, CHole), 140.9 (d, $^1J_{C,F}$ = 247 Hz, 1C, CF), 136.6 (1C, Cq), 131.9 (1C, Cq), 131.8 (2C, 2 x CHar), 131.7 (1C, CHar), 127.9 (2C, 2 x CHar), 124.6 (1C, CHar), 120.3 (1C, Cq), 116.7 (d, $^1J_{C,F}$ = 24 Hz, 1C, Cq), 115.7 (d, $^2J_{C,F}$ = 21 Hz, 1C, CHar), 104.6 (1C, CN).

(2*E*)-2-cyano-3-[4'-(trifluoromethyl)[1,1'-biphenyl]-4-yl]prop-2-enoic acid (53)

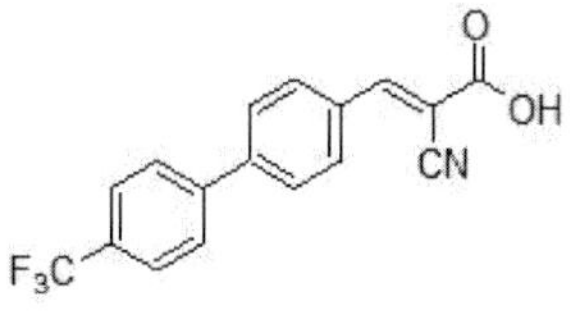

White crystals, 91% yield, Melting Point = 232.9 °C UV-Vis (MeOH: H_2O, 8:2,v/v):λ_{max} = 318 nm MALDI-MS: $[M+H]^+$ = 318.17 Da ^{1}H-NMR (500 MHz, DMSO-d6): δ [ppm] = 8.42 (s, 1H, CHole), 8.19 (d,$^3J_{H,H}$ = 8.20 Hz, 2H, 2 x CHar), 8.02 (d, $^3J_{H,H}$ = 8.20

Hz, 2H, 2 x CHar), 7.99 (d, $^3J_{H,H}$ = 8.20 Hz, 2H, 2 x CHar), .87 (d, $^3J_{H,H}$ = 8.20 Hz, 2H, 2 x CHar). ^{13}C-NMR (125 MHz, DMSO-d6): δ [ppm] = 163.1 (1C, CO), 153.5 (1C, CHole), 142.5 (1C, Cq), 131.4 (1C, Cq), 131.3 (2C, 2 x CHar), 128.5 (1C, Cq), 127.8 (4C, 4 x CHar), 127.7 (2C, 2 x CHar), 125.8 (1C, Cq), 124.2 (q, $^1J_{C,F}$ = 271 Hz, 1C, CF_3), 116.1 (1C, Cq), 103.9 (1C, CN).

(2*E*)-2-cyano-3-[2'-(trifluoromethyl)[1,1'-biphenyl]-4-yl]prop-2-enoic acid (54)

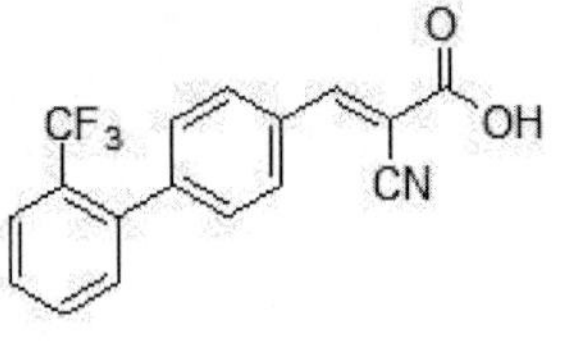

White crystals, 49% yield, Melting Point = 91.0 °C UV-Vis (MeOH: H_2O, 8:2,v/v): λ_{max} = 302 nm MALDI-MS: $[M+H]^+$ = 318.08 Da ^{1}H-NMR (500 MHz, DMSO-d6):δ [ppm] = 8.18 (s, 1H, CHole), 8.03 (d, $^3J_{H,H}$ = 8.20 Hz, 2H, 2 x CHar), 7.87 (d, $^3J_{H,H}$ = 7.90 Hz, 1H, CHar), 7.77-7.74 (m, 1H, CHar), 7.66 (t, $^3J_{H,H}$ = 7.60 Hz, 1H, CHar), 7.49 (d, $^3J_{H,H}$ = 8.20 Hz, 2H, 2 x CHar), 7.46 (d, $^3J_{H,H}$ = 7.60 Hz, 1H, CHar). ^{13}C-NMR (125 MHz, DMSO-d6): δ [ppm] = 162.5 (1C, CO), 148.5 (1C, CHole), 139.6 (1C, Cq), 132.3 (1C, CHar), 131.9 (1C, Cq), 131.8 (1C, CHar), 129.4 (4C, 4 x CHar), 128.4 (1C, CHar), 126.7 (1C, CHar), 126.5 (1C, Cq), 126.1 (1C, Cq), 121.5 (q, $^1J_{C,F}$ = 269 Hz, 1C, CF_3), 116.3 (1C, Cq), 106.9 (1C, CN).

(2*E*)-2-cyano-3-[4'-(trifluoromethyl)[1,1'-biphenyl]-3-yl]prop-2-enoic acid (55)

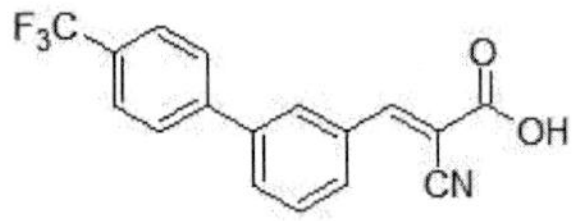

White crystals, 63% yield, Melting Point = 193.8 °C UV-Vis (MeOH: H_2O, 8:2,v/v):λ_{max} = 257 nm MALDI-MS: $[M+H]^+$ = 318.33 Da ^{1}H-NMR (500 MHz, DMSO-d6): δ [ppm] = 8.45 (s, 1H, CHar), 8.39 (s, 1H, CHole), 8.11 (d, $^3J_{H,H}$ = 7.30 Hz, 1H, CHar), 7.99 (d, $^3J_{H,H}$ = 7.60 Hz, 1H, CHar), 7.96 (d, $^3J_{H,H}$ = 7.90 Hz, 2H, 2 x CHar), 7.88 (d, $^3J_{H,H}$ = 7.90 Hz, 2H, 2 x CHar), 7.73 (t, $^3J_{H,H}$ = 7.60 Hz, 1H, CHar). ^{13}C-NMR (125 MHz, DMSO-d6): δ [ppm] = 163.6 (1C, CO), 154.7 (1C, CHole), 143.4 (1C, Cq), 139.9 (1C, Cq), 132.8 (1C, Cq), 131.9 (1C, CHar), 130.6 (1C, CHar), 130.4 (1C, CHar), 130.1 (1C, CHar), 128.1 (4C, 4 x CHar), 126.4 (1C, Cq), 122.7 (q, $^1J_{C,F}$ = 269 Hz, 1C, CF_3), 116.5 (1C, Cq), 105.2 (1C, CN).

(2*E*)-2-cyano-3-[2'-(trifluoromethyl)[1,1'-biphenyl]-3-yl]prop-2-enoic acid (56)

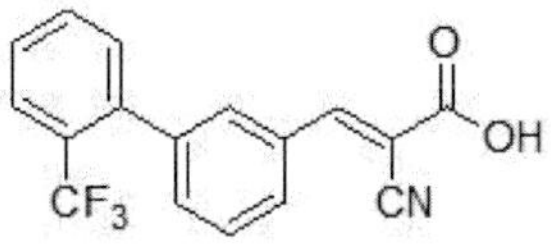

White crystals, 65% yield, Melting Point = 169.2 °C UV-Vis (MeOH: H_2O, 8:2,v/v):λ_{max} = 270 nm MALDI-MS: $[M+H]^+$ = 318.11 Da ^{1}H-NMR (500 MHz, DMSO-d6): δ [ppm] = 8.00 (d, $^3J_{H,H}$ = 6.90 Hz, 1H, CHar), 7.87-7.84 (m, 3H, 2 x CHar, CHole), 7.76-7.73 (m, 1H, CHar), 7.67-7.65 (m, 1H, CHar), 7.60-7.56 (m, 2H, 2 x CHar), 7.44 (d, 3J H,H = 7.60 Hz, 1H, CHar). ^{13}C-NMR (125 MHz, DMSO-d6): δ [ppm] = 162.9 (1C, CO), 153.6 (1C, CHole), 140.0 (1C, Cq), 133.1 (1C, CHar), 132.4 (1C, CHar), 131.9 (1C, CHar), 131.2 (1C, Cq), 130.4 (1C, CHar), 130.0 (1C, CHar), 128.9 (1C, CHar), 128.5 (1C, CHar), 126.7 (1C, Cq), 126.1 (1C, Cq), 122.7 (q, $^1J_{C,F}$ = 277 Hz, 1C, CF_3), 115.8 (1C, Cq), 102.3 (1C, CN).

(2*E*)-2-cyano-3-[4'-(trifluoromethyl)[1,1'-biphenyl]-2-yl]prop-2-enoic acid (57)

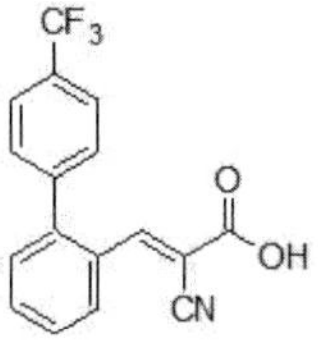

Yellow solid, 72% yield, Melting Point = 226.3 °C UV-Vis (MeOH: H_2O, 8:2,v/v):λ_{max} = 294 nm MALDI-MS: $[M+H]^+$ = 318.02 Da

^{1}H-NMR (500 MHz, DMSO-d6): δ [ppm] = 8.16 (d, $^3J_{H,H}$ = 7.60 Hz, 1H, CHar), 8.03 (s, 1H, CHole), 7.88 (d, $^3J_{H,H}$ = 7.90 Hz, 2H, 2 x CHar), 7.75-7.72 (m, 1H, CHar), 7.70-7.67 (m, 1H, CHar), 7.62 (d, $^3J_{H,H}$ = 7.60 Hz, 3H, 3 x CHar). ^{13}C-NMR (125 MHz, DMSO-d6): δ [ppm] = 162.6 (1C, CO), 153.0 (1C, CHole), 142.7 (1C, Cq), 141.6 (1C, Cq), 132.2 (1C, CHar), 130.6 (2C, 2 x CHar), 130.4 (1C, CHar), 129.8 (1C, Cq), 128.6 (1C, CHar), 125.3 (1C, Cq), 123.4 (q, $^1J_{C,F}$ = 271 Hz, 1C, CF_3), 109.2 (1C, Cq), 105.1 (1C, CN).

(2*E*)-3-[3',5'-bis(trifluoromethyl)[1,1'-biphenyl]-4-yl]-2-cyanoprop-2-enamide (58)

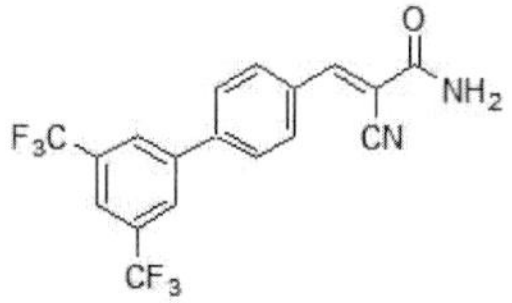

White crystals, 89% yield, Melting Point = 217.2 °C UV-Vis (MeOH: H_2O, 8:2,v/v):λ_{max} = 321 nm MALDI-MS: $[M+H]^+$ = 385.42 Da ^{1}H-NMR (500 MHz, DMSO-d6): δ [ppm] = 8.46 (s,

2H, 2 x CHar), 8.26 (s, 1H, CHar), 8.17 (s, 1H, CHole), 8.13-8.08 (m, 4H, 4 x CHar), 7.98 (s, br, 1H, NH), 7.83 (s, br, 1H, NH). ^{13}C-NMR (125 MHz, DMSO-d6): δ [ppm] = 162.5 (1C, CO), 149.6 (1C, CHole), 140.0 (1C, Cq), 131.5 (1C, CHar), 130.8 (2C, 2 x Cq), 130.6 (2C, 2 x CHar), 129.9 (1C, Cq), 128.1 (4C, 4 x CHar), 127.6 (1C, Cq), 124.3 (q, $^{1}J_{C,F}$ = 271 Hz, 2C, 2 x CF_3), 116.1 (1C, Cq), 107.2 (1C, CN).

(2*E*)-3-[3',5'-bis(trifluoromethyl)[1,1'-biphenyl]-4-yl]-2-cyanoprop-2-enoic acid (59)

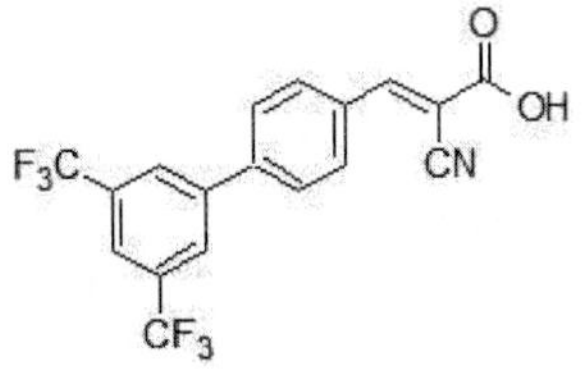

White crystals, 16% yield, Melt-ing Point = 227.5 °C UV-Vis (MeOH: H_2O, 8:2,v/v):λ_{max} = 312 nm MALDI-MS: $[M+H]^+$ = 386.33 Da ^{1}H-NMR (500 MHz, DMSO-d6): δ [ppm] = 8.46 (s, 2H, 2 x CHar), 8.43 (s, 1H, CHar), 8.20-8.17 (m, 3H, 2 x CHar, CHole), 8.13 (d, $^{3}J_{H,H}$ = 7.95 Hz, 2H, 2 x CHar). ^{13}C-NMR (125 MHz, DMSO-d6): δ [ppm] = 163.0 (1C, CO), 153.1 (1C, CHole), 141.1 (1C, Cq), 140.7 (1C, Cq), 131.9 (2C, 2 x Cq), 131.2 (2C, 2 x CHar), 130.8 (1C, CHar), 128.1 (4C, 4 x CHar), 127.7 (1C, Cq), 123.1 (q, $^{1}J_{C,F}$ = 271 Hz, 1C, CF_3), 116.1 (1C, Cq), 104.6 (1C, CN).

6.1.2.2 HCCA Derivatives

(2*E*)-2-cyano-3-[4-(diethoxymethyl)phenyl]prop-2-enamide

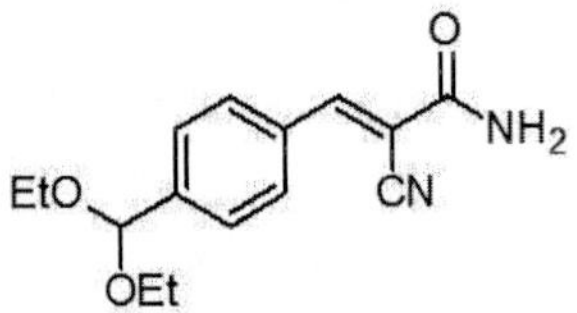

White crystals, 33% yield, Melting Point = 143.7 °C UV-Vis (MeOH: H_2O, 8:2,v/v):λ_{max} = 305 nm ^{1}H-NMR (500 MHz, DMSO-d6): δ [ppm] = 8.18 (s, 1H, CHole), 8.08 (s, br, 1H, NH), 7.94 (d, $^{3}J_{H,H}$ = 7.60 Hz, 2H, 2 x CHar), 7.78 (s, br, 1H, NH), 7.57 (d, $^{3}J_{H,H}$ = 7.90 Hz, 2H, 2 x CHar), 5.56 (s,

1H, CH), 3.58- 3.50 (m, 4H, 2 x CH2), 1.16 (t, $^3J_{H,H}$ = 7.00 Hz, 6H, 2 x CH_3). ^{13}C-NMR (125 MHz, DMSO-d6): δ [ppm] = 162.2 (1C, CO), 150.0 (1C, CHole), 143.2 (1C, CHar), 130.7 (1C, Cq), 130.3 (1C, Cq), 129.8 (2C, 2 x CHar), 129.1 (2C, 2 x CHar), 116.3 (1C, Cq), 106.7 (1C, CN) 100.2 (1C, CH), 60.9 (2C, 2 x CH_2), 15.0 (2C, 2 x CH_3).

(2*E*)-2-cyano-3-[4-(trifluoromethyl)phenyl]prop-2-enoic acid

White crystals, 76% yield, Melting Point = 161.2 °C UV-Vis (MeOH: H_2O, 8:2,v/v):λ_{max} = 280 nm ^{1}H-NMR (500 MHz, DMSO-d6): δ [ppm] = 8.45 (s, 1H, CHole), 8.19 (d, $^3J_{H,H}$ = 8.25 Hz, 2H, 2 x CHar), 7.97 (d, $^3J_{H,H}$ = 8.25 Hz, 2H, 2 x CHar). ^{13}C-NMR (125 MHz, DMSO-d6): δ [ppm] = 162.6 (1C, CO), 152.5 (1C, CHole), 130.9 (2 C, 2 x CHar), 125.9 (1C, Cq), 124.1 (q, $^1J_{C,F}$ = 271 Hz, 1C, CF_3), 115.5 (1C, Cq), 108.1 (1C, CN).

(2*E*)-2-cyano-3-[4-(1,1,2,2-tetrafluoroethoxy)phenyl]prop-2-enoic acid

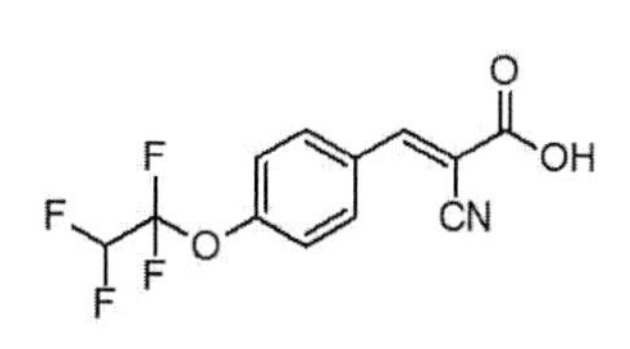

White crystals, 74% yield, Melting Point = 178.5 °C UV-Vis (MeOH: H_2O, 8:2,v/v):λ_{max} = 292 nm MALDI-MS: $[M+H]^+$ = 290.32 Da ^{1}H-NMR (500 MHz, DMSO-d6): δ [ppm] = 14.09 (s, 1H, COOH), 8.38 (s, 1H, CHole) 8.15 (d, $^3J_{H,H}$ = 8.55 Hz, 2H, 2 x CHar), 7.52 (d, $^3J_{H,H}$ = 8.45 Hz, 2H, 2 x CHar), 6.87 (1H, $^2J_{H,F}$ = 51.80 Hz, CH). ^{13}C-NMR (125 MHz, DMSO-d6): δ [ppm] = 162.9 (1C, CO), 152.8 (1C, CHole), 150.9 (1C, Cq), 132. 6 ((2 C, 2 x CHar), 130.0 (1C, Cq), 121.8 ((2 C, 2 x CHar), 115.7 (1C, Cq), 104.5 (1C, CN).

(2*E*)-3-(4-butoxyphenyl)-2-cyanoprop-2-enoic acid

Pale yellow crystals, 81% yield, Melting Point = 160.1 °C UV-Vis (MeOH: H_2O, 8:2,v/v):λ_{max} = 326 nm ^{1}H-NMR (500 MHz, DMSO-d6): δ [ppm] = 13.07 (s, 1H, COOH), 8.25 (d, $^3J_{H,H}$ = 8.70 Hz, 2H, 2 x CHar), 7.14 (d, $^3J_{H,H}$ = 8.70 Hz, 2H, 2 x

CHar), 4.09 (t, $^{3}J_{H,H}$ = 6.30 Hz, 2H, 2 x CH_2), 1.75-1.69 (m, 2H, CH_2), 1.48-1.40 (m, 2H, CH_2), 0.94 (t, $^{3}J_{H,H}$ = 7.35 Hz, 3H, CH_3).

(2*E*)-2-cyano-3-3-[(2-methylprop-2-en-1-yl)oxy]phenylprop-2-enoic acid

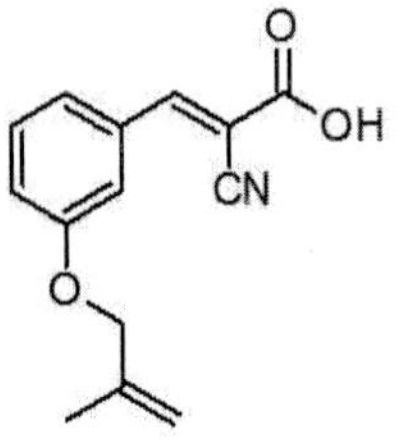

Pale yellow crystals, 76% yield, Melting Point = 120.3 °C UV-Vis (MeOH: H_2O, 8:2,v/v):λ_{max} = 291 nm MALDI-MS: $[M+H]^+$ = 244.27 Da ^{1}H-NMR (500 MHz, DMSO-d6): δ [ppm] = 8.31 (s, 1H, CHole), 7.64 (s, 1H, CHar), 7.62 (s, 1H, CHar), 7.49 (t, $^{3}J_{H,H}$ = 7.85 Hz, 1H, CHar), 7.22 (d, $^{3}J_{H,H}$ = 8.45 Hz, 1H, CHar), 5.07 (s, 1H, CH), 4.98 (s, 1H, CH), 4.54 (s, 2H, CH_2), 1.80 (s, 3H, CH_3). ^{13}C-NMR (125 MHz, DMSO-d6): δ [ppm] = 163.1 (1C, CO), 158.2 (1C, Cq), 154.4 (1C, CHole), 140.4 (1C, Cq), 132.9 (1C, Cq), 130.2 (1C, CHar), 123.2 (1C, CHar), 120.0 (1C, CHar), 115.8 (1C, CHar), 112.6 (1C, CH_2), 103.9 (1C, CN), 71.2 (1C, CH_2), 19.3 (1C, CH_3).

(2*E*)-2-cyano-3-(4-ethoxy-3-methoxyphenyl)prop-2-enoic acid

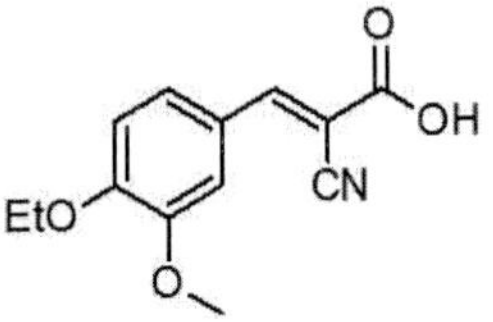

Yellow crystals, 90% yield, Melting Point = 215.21 °C UV-Vis (MeOH: H_2O, 8:2,v/v):λ_{max} = 241 nm ^{1}H-NMR (500 MHz, DMSO-d6): δ [ppm] = 13.75 (s, 1H, COOH), 8.27 (s, 1H, CHole), 7.76 (s, 1H, CHar), 7.68 (d, $^{3}J_{H,H}$ = 8.25 Hz, 1H, CHar), 7.16 (d, $^{3}J_{H,H}$ = 8.55 Hz, 1H, CHar), 4.14 (q, $^{3}J_{H,H}$ = 6.65 Hz, 2H, CH_3), 3.81 (s, 3H, CH_3), 1.36 (t, $^{3}J_{H,H}$ = 6.65Hz, 1H, CH3). ^{13}C-NMR (125 MHz, DMSO-d6): δ [ppm] = 163.9 (1C, CO), 154.0 (1C, CHole), 152.2 (1C, Cq), 148.7 (1C, Cq), 126.4 (1C, CHar), 124.0 (1C, Cq), 117.0 (1C, Cq), 112.8 (1C, CHar), 112.4 (1C, CHar), 99.7 (1C, CN), 64.1 (1C, CH_2), 55.5 (1C, CH_3), 14.3 (1C, CH_3).

(2*E*)-3-(4-chlorophenyl)-2-cyanoprop-2-enehydrazide

White crystals, 71% yield, Melting Point = 204.3 °C UV-Vis (MeOH: H_2O, 8:2,v/v):λ_{max} = 286 nm

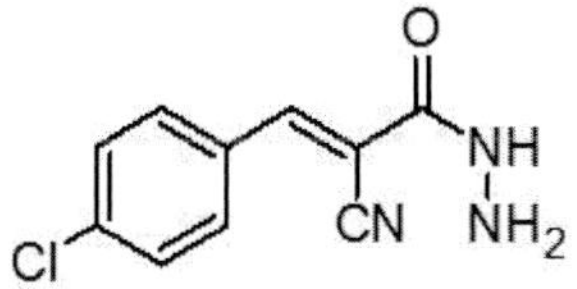

^{1}H-NMR (500 MHz, DMSO-d6): δ [ppm] = 11.87 (s, 1H, NH), 8.00 (s, 1H, CHole), 7.74 (d, $^3J_{H,H}$ = 8.25 Hz, 1H, CHar), 7.51 (d, $^3J_{H,H}$ = 8.25 Hz, 1H, CHar), 4.23 (s, 2H, NH_2). ^{13}C-NMR (125 MHz, DMSO-d6): δ [ppm] = 165.0 (1C, CO), 146.2 (1C, CHole), 143.3 (1C, CHole), 134.6 (1C, Cq), 132.6 (1C, Cq), 128.8 (2C, 2 x CHar), 128.6 (2C, 2 x CHar), 115.9 (1C, CN).

(2*E*)-3-(4-butoxyphenyl)-2-cyanoprop-2-enamide

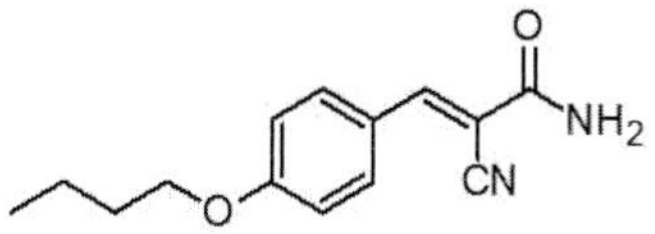

White crystals, 86% yield, Melting Point = 133.3 °C UV-Vis (MeOH: H_2O, 8:2,v/v):λ_{max} = 338 nm ^{1}H-NMR (500 MHz, DMSO-d6): δ [ppm] = 8.09 (s, 1H, CHole), 7.94 (d, $^3J_{H,H}$ = 8.55 Hz, 2H, 2 x CHar), 7.79 (s, br, 1H, NH), 7.66 (s, br, 1H, NH), 7.12 (d, $^3J_{H,H}$ = 8.55 Hz, 2H, 2 x CHar), 4.08 (t, $^3J_{H,H}$ = 6.35 Hz, 2H, CH_2), 1.75-1.69 (m, 2H, CH_2), 1.46-1.42 (m, 2H, CH_2), 0.94 (t, $^3J_{H,H}$ = 6.35 Hz, 3H, CH_3). ^{13}C-NMR (125 MHz, DMSO-d6): δ [ppm] = 163.2 (1C, CO), 162.2 (1C, Cq), 150.4 (1C, CHole), 132.2 (2C, 2 x CHar), 124.1 (1C, Cq), 115.1 (2C, 2 x CHar), 102.6 (1C, CN), 67.8 (1C, CH_2), 30.6 (1C, CH_2), 18.3 (1C, CH_2), 13.6 (1C, CH_3).

(2*E*)-3-[3-(benzyloxy)-4-methoxyphenyl]-2-cyanoprop-2-enamide

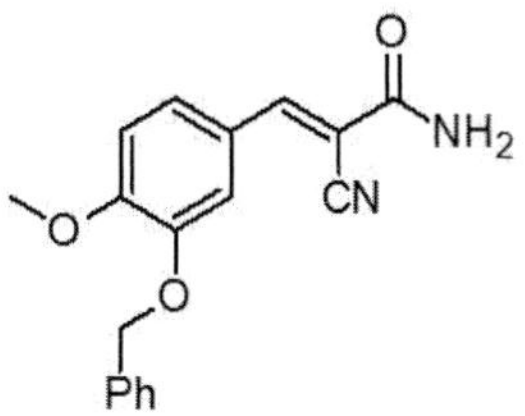

White crystals, 86% yield, Melting Point = 182.9 °C UV-Vis (MeOH: H_2O, 8:2,v/v):λ_{max} = 354 nm ^{1}H-NMR (500 MHz, DMSO-d6): δ [ppm] = 8.08 (s, 1H, CHole), 7.78 (s, br, 1H, NH), 7.74 (s, 1H, CHar), 7.67 (s, br, 1H, NH), 7.61 (d, $^3J_{H,H}$ = 8.55 Hz, 1H, CHar), 7.48-7.34 (m, 5H, 5 x CHar), 7.19 (d, $^3J_{H,H}$ = 8.85 Hz, 1H, CHar), 5.11 (s,

2H, CH_2), 3.87 8s, 3H, CH_3). ^{13}C-NMR (125 MHz, DMSO-d6): δ [ppm] = 163.1 (1C, CO), 152.9 (1C, Cq), 150.5 (1C, CHar), 147.5 (1C, CHar), 136.3 (1C, Cq), 128.4 (2C, 2 x CHar), 127.9 (2C, 2 x CHar), 125.5 (1C, CHar), 124.3 (1C, Cq), 117.0 (1C, Cq), 114.2 (1C, CHar), 112.1 (1C, CHar), 102.7 (1C, CN), 69.8 (1C, CH_2), 55.8 (1C, CH_3).

(2*E*)-2-cyano-3-(3-ethoxy-4-propoxyphenyl)prop-2-enamide

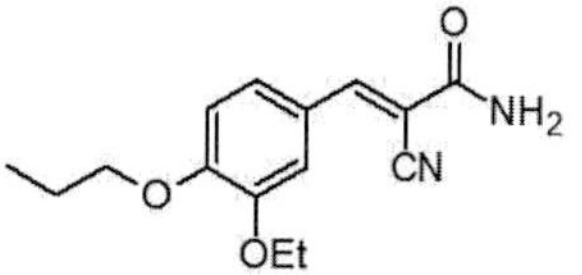

O NH2 CN O OEt

White crystals, 82% yield, Melting Point = 160.7 °C UV-Vis (MeOH: H_2O, 8:2,v/v):λ_{max} = 356 nm ^{1}H-NMR (500 MHz, DMSO-d6): δ [ppm] = 8.08 (s, 1H, CHole), 7.79 (s, br, 1H, NH), 7.65 (s, br, 2H, CHar, NH), 7.52 (d, $^3J_{H,H}$ = 8.30 Hz, 1H, CHar), 7.14 (d, $^3J_{H,H}$ = 8.30 Hz, 1H, CHar), 4.08-4.02 (m, 4H, 2 x CH_2), 1.78-1.73 (m, 2H, CH_2), 1.35 (t, $^3J_{H,H}$ = 7.00 Hz, 3H, CH_3), 0.98 (t, $^3J_{H,H}$ = 7.30 Hz, 3H, CH_3). ^{13}C-NMR (125 MHz, DMSO-d6): δ [ppm] = 163.1 (1C, CO), 152.4 (1C, Cq), 150.4 (1C, CHar), 134.8 (1C, Cq), 125.4 (1C, CHar), 113.8 (1C, CHar), 112.3 (1C, CHar), 102.9 (1C, CN), 69.8 (1C, CH_2), 63.8 (1C, CH_2), 21.8 (1C, CH_2), 14.6 (1C, CH_3), 10.6 (1C, CH_3).

(2*E*)-2-cyano-3-[4-(trifluoromethoxy)phenyl]prop-2-enoic acid

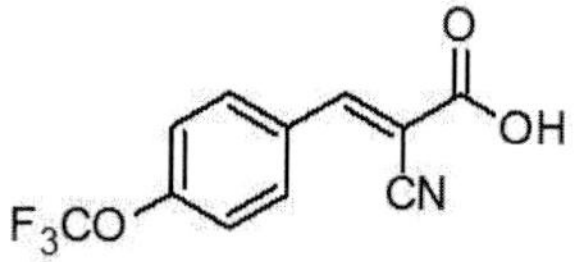

O OH CN F_3CO

White crystals, 72% yield, Melting Point = 147.1 °C UV-Vis (MeOH: H_2O, 8:2,v/v):λ_{max} = 289 nm ^{1}H-NMR (500 MHz, DMSO-d6): δ [ppm] = 8.40 (s, 1H, CHole), 8.16 (d, $^3J_{H,H}$ = 8.55 Hz, 2H, 2 x CHar), 7.60 (d, $^3J_{H,H}$ = 8.55 Hz, 2H, 2 x CHar). ^{13}C-NMR (125 MHz, DMSO-d6): δ [ppm] = 163.1 (1C, CO), 152.9 (1C, CHole), 151.0 (1C, Cq), 132.9 (2C, 2 x CHar), 130.9 (1C, Cq), 125.5 (2C, 2 x CHar), 120.8 (1C, Cq), 118.8 (q, $^1J_{C,F}$ = 271 Hz, 1C, CF_3), 104.9 (1C, CN).

(2E)-3-[3-(benzyloxy)-4-methoxyphenyl]-2-cyanoprop-2-enoic acid

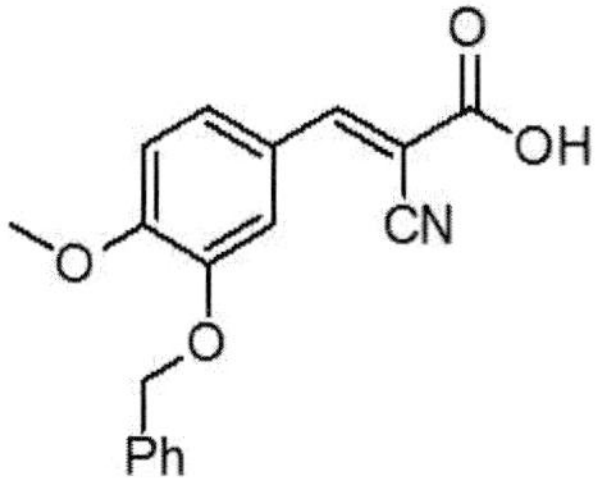

Pale yellow crystals, 73% yield, Melting Point = 202.1.1 °C UV-Vis (MeOH: H_2O, 8:2,v/v):λ_{max} = 337 nm ^{1}H-NMR (500 MHz, DMSO-d6): δ [ppm] = 8.23 (s, 1H, CHole), 7.84 (s, 1H, CHar), 7.74 (d, $^3J_{H,H}$ = 8.55 Hz, 1H, CHar), 7.48-7.34 (m, 5H, 5 x CHar), 7.20 (d, $^3J_{H,H}$ = 8.85 Hz, 1H, CHar), 5.14 (s, 2H, CH_2), 3.91 8s, 3H, CH_3) ^{13}C-NMR (125 MHz, DMSO-d6): δ [ppm] =164.3 (1C, CO), 154.6 (1C, Cq), 154.0 (1C, CHar), 147.5 (1C, CHar), 136.8 (1C, Cq), 128.9 (2C, 2 x CHar), 128.6 (2C, 2 x CHar), 126.9 (1C, CHar), 124.6 (1C, Cq), 117.2 (1C, Cq), 115.2 (1C, CHar), 112.7 (1C, CHar), 100.3 (1C, CN), 70.4 (1C, CH_2), 56.4 (1C, CH_3).

(2*E*)-2-cyano-3-4-[(trifluoromethyl)sulfanyl]phenylprop-2-enoic acid

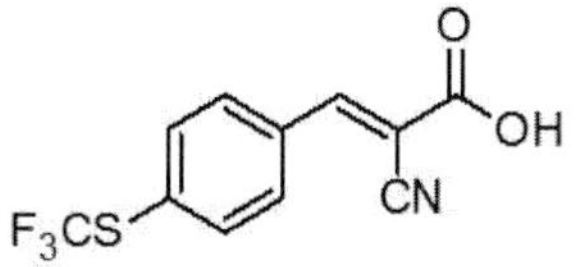

Pale yellow crystals, 66% yield, Melting Point = 188.9 °C UV-Vis (MeOH: H_2O, 8:2,v/v):λ_{max} = 294 nm ^{1}H-NMR (500 MHz, DMSO-d6): δ [ppm] = 8.34 (s, 1H, CHole), 8.11 (d, 3JH,H = 8.55 Hz, 2H, 2 x CHar), 7.91 (d, $^3J_{H,H}$= 8.55 Hz, 2H, 2 x CHar)

^{13}C-NMR (125 MHz, DMSO-d6): δ [ppm] =162.8 (1C, CO), 153.2 (1C, CHole), 136.4 (2C, 2 x CHar), 133.7 (1C, Cq), 131.4 (2C, 2 x CHar), 120.8 (1C, Cq), 127.5 (q, $^1J_{C,F}$ = 271 Hz, 1C, CF_3), 115.8 (1C, Cq), 106.6 (1C, CN).

(2*E*)-2-cyano-3-[4-(1,1,2,2-tetrafluoroethoxy)phenyl]prop-2-enamide

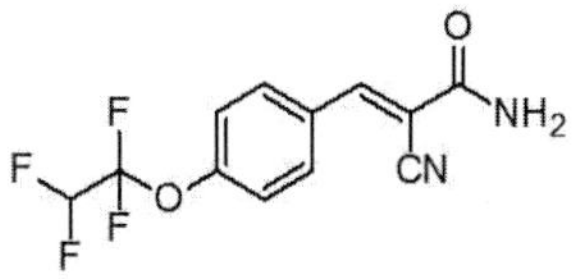

White crystals, 72% yield, Melting Point = 121.3 °C UV-Vis (MeOH: H_2O, 8:2,v/v):λ_{max} = 300 nm ^{1}H-NMR (500 MHz, DMSO-d6): δ [ppm] = 8.21 (s, 1H, CHole) 8.04 (d, $^3J_{H,H}$ = 8.55 Hz, 2H, 2 x CHar), 7.94 (s, br, 1H, NH), 7.81 (s, br, 1H, NH), 7.50 (d, $^3J_{H,H}$ = 8.45 Hz, 2H, 2 x CHar), 6.87 (1H, $^2J_{H,F}$ = 52.10 Hz, CH) ^{13}C-NMR (125 MHz, DMSO-d6): δ [ppm] =162.6 (1C, CO),

150.4 (1C, Cq), 149.2 (1C, CHole), 131.9 ((2 C, 2 x CHar), 130.6 (1C, Cq), 122.0 (2 C, 2 x CHar), 116.1 (1C, Cq), 107.4 (1C, CN).

(2*E*)-4-(4-decyloxy)-2-cyanoprop-2-enamide

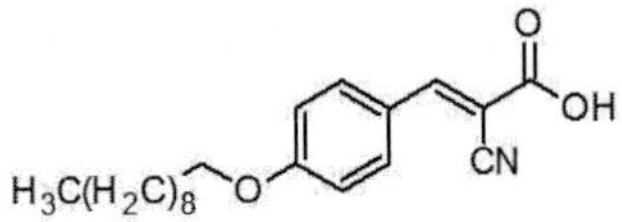

Pale yellow crystals, 87% yield, Melting Point = 133.1 °C UV-Vis (MeOH: H_2O, 8:2,v/v):λ_{max} = 326 nm ^{1}H-NMR (500 MHz, DMSO-d6): δ [ppm] = 13.93 (s, 1H, COOH), 8.25 (s, 1H, CHole), 7.05 (d, $^3J_{H,H}$ = 8.55 Hz, 2H, 2 x CHar), 7.12 (d, $^3J_{H,H}$ = 8.55 Hz, 2H, 2 x CHar), 4.08 (t, $^3J_{H,H}$ = 6.35 Hz, 2H, CH_2), 1.75-1.69 (m, 2H, CH_2), 1.44-1.43 (m, 2H, CH_2), 1.30-1.18 (m, 12H, 6 x CH_2), 0.86 (t, $^3J_{H,H}$ = 6.35 Hz, 3H, CH_3) ^{13}C-NMR (125 MHz, DMSO-d6): δ [ppm] =164.2 (1C, CO), 163.4 (1C, Cq), 154.4 (1C, CHole), 133.9 (2C, 2 x CHar), 124.6 (1C, Cq), 117.2 (1C, Cq), 116.0 (2C, 2 x CHar), 99.9 (1C, CN), 68.5 (1C, CH_2), 31.6 (1C, CH_2), 29.5 (1C, CH_2), 29.4 (1C, CH_2), 29.2 (1C, CH_2), 29.1 (1C, CH_2), 28.9 (1C, CH_2), 25.8 (1C, CH_2), 22.5 (1C, CH_2), 14.4 (1C, CH_3)

(2*E*)-2-cyano-3-[4-(methylsulfanyl)phenyl]prop-2-enamide

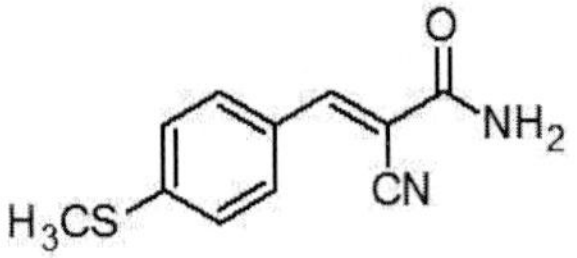

Pale yellow crystals, 74% yield, Melting Point = 187.2 °C UV-Vis (MeOH: H_2O, 8:2,v/v):λ_{max} = 357 nm ^{1}H-NMR (500 MHz, DMSO-d6): δ [ppm] = 8.12 (s, 1H, CHole), 7.89 (d, $^3J_{H,H}$ = 8.55 Hz, 2H, 2 x CHar), 7.86 (s, br, 1H, NH), 7.73 (s, br, 1H, NH), 7.43 (d, $^3J_{H,H}$ = 8.55 Hz, 2H, 2 x CHar), 2.55 (s, 3H, CH_3) ^{13}C-NMR (125 MHz, DMSO-d6): δ [ppm] =162.9 (1C, CO), 149.8 (1C, CHole), 145.1 (1C, CHar), 130.4 (2C, 2 x CHar), 130.1 (1C, CHar), 127.8 (1C, Cq), 125.4 (2C, 2 x CHar), 125.1 (1C, CHar), 116.6 (1C, Cq), 104.6 (1C, CN), 13.8 (1C, CH_3).

(2*E*)-3-(4-tert-butoxyphenyl)-2-cyanoprop-2-enamide

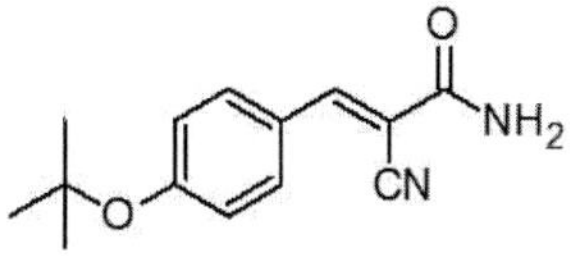

White crystals, 86% yield, Melting Point = 134.2 °C UV-Vis (MeOH: H2O, 8:2,v/v):λ_{max} = 327 nm MALDI-MS: $[M+H]^+$ = 245.56 Da ^{1}H-NMR (500 MHz, DMSO-d6): δ [ppm] = 8.12 (s, 1H, CHole), 7.91 (d, $^3J_{H,H}$ = 8.55 Hz, 2H, 2 x CHar), 7.83 (s, br, 1H, NH), 7.70 (s, br, 1H, NH), 7.16 (d, $^3J_{H,H}$ = 8.55 Hz, 2H, 2 x CHar), 1.40 (s, 9H, 3 x CH_3) ^{13}C-NMR (125 MHz, DMSO-d6): δ [ppm] =162.9 (1C, CO), 159.3 (1C, Cq), 150.0 (1C, CHole), 131.7 (2C, 2 x CHar), 125.9 (1C, Cq), 122.0 (2C, 2 x CHar), 125.1 (1C, CHar), 117.1 (1C, Cq), 103.5 (1C, CN), 28.4 (3C, 3 x CH_3).

(2*E*)-3-(4-tert-butoxyphenyl)-2-cyanoprop-2-enoic acid

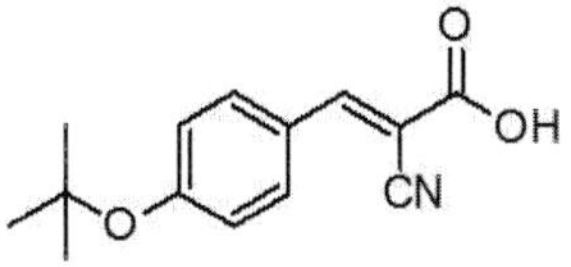

Pale yellow crystals, 52% yield, Melting Point = 226.6 °C UV-Vis (MeOH: H_2O, 8:2,v/v):λ_{max} = 327 nm MALDI-MS: $[M+H]^+$ = 246.73 Da ^{1}H-NMR (500 MHz, DMSO-d6): δ [ppm] = 8.26 (s, 1H, CHole), 8.02 (d, $^3J_{H,H}$ = 8.55 Hz, 2H, 2 x CHar), 7.17 (d, $^3J_{H,H}$ = 8.55 Hz, 2H, 2 x CHar), 1.40 (s, 9H, 3 x CH_3) ^{13}C-NMR (125 MHz, DMSO-d6): δ [ppm] =163.4 (1C, CO), 160.1 (1C, Cq), 153.5 (1C, CHole), 132.7 (2C, 2 x CHar), 125.3 (1C, Cq), 121.8 (2C, 2 x CHar), 125.1 (1C, CHar), 116.6 (1C, Cq), 100.7 (1C, CN), 79.9 (1C, Cq), 28.4 (3C, 3 x CH_3).

(2*E*)-2-cyano-3-(3,4-diethoxyphenyl)prop-2-enamide

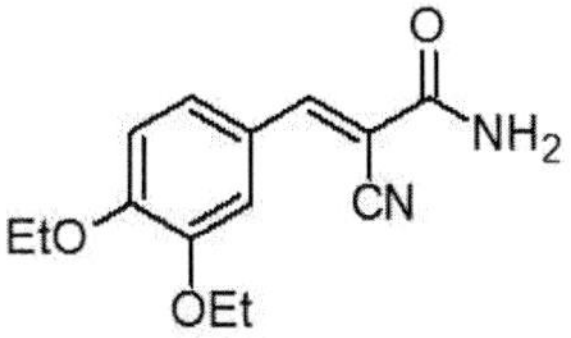

White crystals, 85% yield, Melting Point = 174.9 °C UV-Vis (MeOH: H2O, 8:2,v/v):λ_{max} = 355 nm ^{1}H-NMR (500 MHz, DMSO-d6): δ [ppm] = 8.08 (s, 1H, CHole), 7.78 (s, br, 1H, NH), 7.66 (s, 2H, CHar, NH), 7.66 (s, br, 1H, NH), 7.22 (d, $^3J_{H,H}$ = 8.55 Hz, 1H, CHar), 7.14 (d, $^3J_{H,H}$ = 8.55 Hz, 1H, CHar), 4.16-4.05 (m, 4H, 2 x CH_2), 1.36 (t, $^3J_{H,H}$ = 6.35 Hz, 3H, CH_3) ^{13}C-NMR (125 MHz, DMSO-d6): δ [ppm] =163.6 (1C, CO), 152.5 (1C, Cq), 151.2 (1C, CHole), 148.2 (1C, Cq), 126.4 (1C, CHar),

125.1 (1C, Cq), 117.9 (1C, Cq), 113.8 (1C, CHar), 113.5 (1C, CHar), 103.0 (1C, CN), 64.8 (1C, CH_2), 64.5 (1C, CH_2), 15.1 (1C, CH_3).

(2*E*)-2-cyano-3-[4-(cyclopentyloxy)phenyl]prop-2-enoic acid

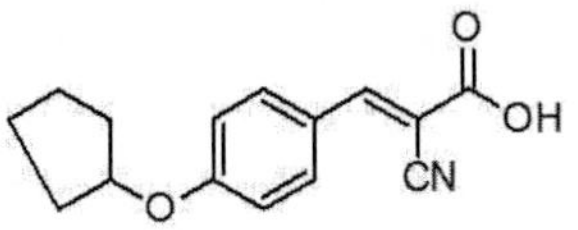

Pale yellow crystals, 77% yield, Melting Point = 179.4 °C UV-Vis (MeOH: H_2O, 8:2,v/v):λ_{max} = 329 nm ^{1}H-NMR (500 MHz, DMSO-d6): δ [ppm] = 13.72 (s, 1H, COOH), 8.25 (s, 1H, CHole), 8.04 (d, $^3J_{H,H}$ = 8.55 Hz, 2H, 2 x CHar), 7.10 (d, $^3J_{H,H}$ = 8.55 Hz, 2H, 2 x CHar), 5.04-4.92 (m, 1H, CH), 2.02-1.92 (m, 2H, CH_2), 1.76.-1-67 (m, 4H, 2 x CH_2), 1.63-1.57 (m, 2H, CH_2) ^{13}C-NMR (125 MHz, DMSO-d6): δ [ppm] =163.6 (1C, CO), 161.7 (1C, Cq), 154.0 (1C, CHole), 133.1 (2C, 2 x CHar), 123.7 (1C, Cq), 116.2 (2C, 2 x CHar), 99.6 (1C, CN), 79.2 (1C, CH), 32.5 (2C, 2 x CH_2), 23.4 (2C, 2 x CH_2).

(2*E*)-2-cyano-3-[4-(cyclopentyloxy)phenyl]prop-2-enamide

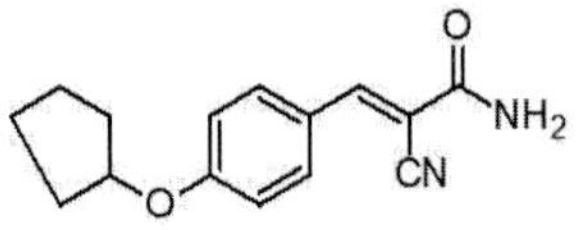

White crystals, 74% yield, Melting Point = 133.0 °C UV-Vis (MeOH: H_2O, 8:2,v/v):λ_{max} = 342 nm ^{1}H-NMR (500 MHz, DMSO-d6): δ [ppm] = 8.09 (s, 1H, CHole), 7.94 (d, $^3J_{H,H}$ = 8.55 Hz, 2H, 2 x CHar), 7.79 (s, 2H, CHar, NH), 7.66 (s, br, 1H, NH), 7.02 (d, $^3J_{H,H}$ = 8.55 Hz, 2H, 2 x CHar), 4.97-4.92 (m, 1H, CH), 2.02-1.92 (m, 2H, CH_2), 1.76.-1-67 (m, 4H, 2 x CH_2), 1.63-1.57 (m, 2H, CH_2) ^{13}C-NMR (125 MHz, DMSO-d6): δ [ppm] =163.2 (1C, CO), 160.9 (1C, Cq), 150.0 (1C, CHole), 132.1 (2C, 2 x CHar), 123.7 (1C, Cq), 116.0 (2C, 2 x CHar), 102.3 (1C, CN), 79.2 (1C, CH), 32.5 (2C, 2 x CH_2), 23.4 (2C, 2 x CH_2).

(2*E*)-2-cyano-3-(4-ethoxy-3-methoxyphenyl)prop-2-enamide

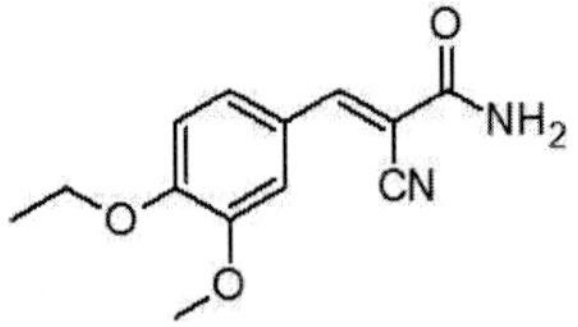

White crystals, 86% yield, Melting Point = 172.3 °C UV-Vis (MeOH: H_2O, 8:2,v/v):λ_{max} = 357 nm ^{1}H-NMR (500 MHz, DMSO-d6): δ [ppm] = 8.09 (s,

1H, CHole), 7.79 (s, 2H, CHar, NH), 7.67 (s, 1H, CHar), 7.55 (d, $^3J_{H,H}$ = 8.55 Hz, 1H, CHar), 7.14 (d, $^3J_{H,H}$ = 8.55 Hz, 1H, CHar), 4.12 (q, $^3J_{H,H}$ = 6.50 Hz, 2H, CH_2), 21.36 (t, $^3J_{H,H}$ = 6.50 Hz, 3H, CH_3) ^{13}C-NMR (125 MHz, DMSO-d6): δ [ppm] =163.2 (1C, CO), 152.0 (1C, Cq), 150.7 (1C, CHole), 125.4 (1C, CHar), 124.4 (1C, Cq), 121.5 (1C, CHar), 102.6 (1C, CN), 64.1 (1C, CH_2), 55.2 (1C, CH_3), 14.2 (1C, CH_3).

(2*E*)-2-cyano-3-4-[(trifluoromethyl)sulfanyl]phenylprop-2-enamide

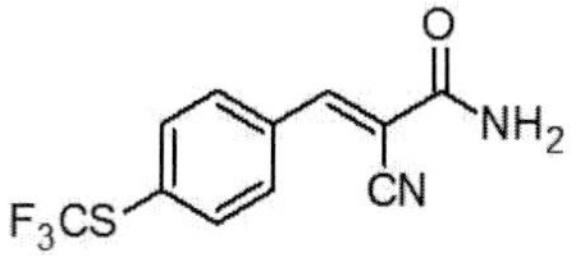

White crystals, 67% yield, Melting Point = 135.0 °C UV-Vis (MeOH: H_2O, 8:2,v/v):λ_{max} = 301 nm ^{1}H-NMR (500 MHz, DMSO-d6): δ [ppm] = 8.24 (s, 1H, CHole), 8.02 (d, $^3J_{H,H}$ = 8.55 Hz, 3H, 2 x CHar, NH), 7.91 (d, $^3J_{H,H}$ = 8.55 Hz, 2H, 2 x CHar), 7.87 (s, br, 1H, NH) ^{13}C-NMR (125 MHz, DMSO-d6): δ [ppm] =162.4 (1C, CO), 149.0 (1C, CHole), 136.4 (2C, 2 x CHar), 134.4 (1C, Cq), 131.1 (2C, 2 x CHar), 120.8 (1C, Cq), 127.5 (q, , $^1J_{C,F}$ = 271 Hz, 1C, CF_3), 115.7 (1C, Cq), 109.6 (1C, CN).

(2*E*)-2-cyano-3-[4-(methylsulfanyl)phenyl]prop-2-enamide

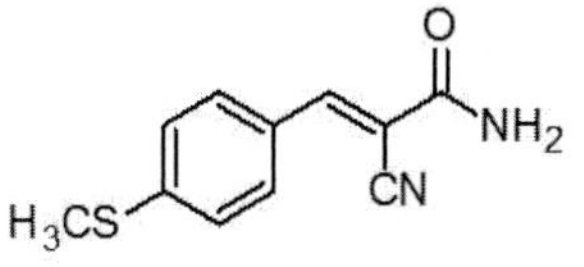

Bright yellow crystals, 77% yield, Melting Point = 244.2 °C UV-Vis (MeOH: H_2O, 8:2,v/v):λ_{max} = 344 nm ^{1}H-NMR (500 MHz, DMSO-d6): δ [ppm] = 8.27 (s, 1H, CHole), 7.99 (d, 3JH,H = 8.55 Hz, 2H, 2 x CHar), 7.43 (d, $^3J_{H,H}$ = 8.55 Hz, 2H, 2 x CHar), 2.55 (s, 3H, CH_3) ^{13}C-NMR (125 MHz, DMSO-d6): δ [ppm] =163.9 (1C, CO), 154.4 (1C, CHole), 146.8 (1C, Cq), 131.6 (2C, 2 x CHar), 128.2 (1C, Cq), 126.1 (2C, 2 x CHar), 117.2 (1C, Cq), 102.1 (1C, CN), 31.5 (1C, CH_3).

(2*E*)-2-cyano-3-3-[(2-methylprop-2-en-1-yl)oxy]phenylprop-2-enamide

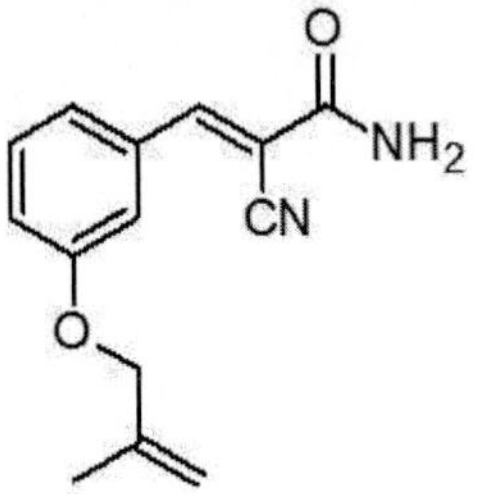

White crystals, 88% yield, Melting Point = 80.6 °C UV-Vis (MeOH: H_2O, 8:2,v/v):λ_{max} = 299 nm ^{1}H-NMR (500 MHz, DMSO-d6): δ [ppm] = 8.13 (s, 1H, CHole), 7.92 8s, br, 1H, NH), 7.78 (s, br, 1H, NH), 7.53-7.45 (m, 3H, 3 x CHar), 7.18 (d, $^3J_{H,H}$ = 8.45 Hz, 1H, CHar), 5.07 (s, 1H, CH), 4.98 (s, 1H, CH), 4.54 (s, 2H, CH_2), 1.77 (s, 3H, CH_3) ^{13}C-NMR (125 MHz, DMSO-d6): δ [ppm] =162.4 (1C, CO), 158.2 (1C, Cq), 150.5 (1C, CHole), 140.4 (1C, Cq), 133.1 (1C, Cq), 130.4 (1C, CHar), 122.2 (1C, CHar), 118.9 (1C, CHar), 115.8 (1C, CHar), 112.6 (1C, CH_2), 103.9 (1C, CN), 71.2 (1C, CH_2), 19.1 (1C, CH_3).

(2E)-2-cyano-3-(3,4-diethoxyphenyl)prop-2-enoic acid

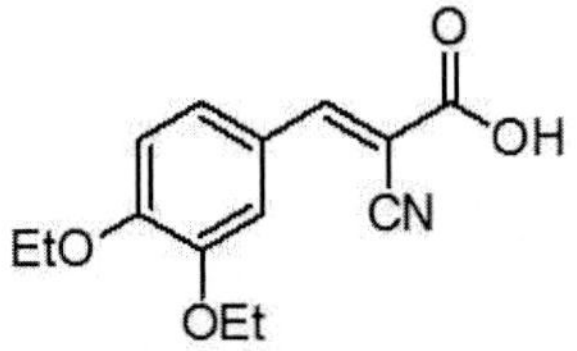

Bright yellow crystals, 86% yield, Melting Point = 116.3 °C UV-Vis (MeOH: H_2O, 8:2,v/v):λ_{max} = 343 nm ^{1}H-NMR (500 MHz, DMSO-d6): δ [ppm] = 13.73 (s, 1H, COOH), 8.24 (s, 1H, CHole), 7.77 (s, 1H, CHar), 7.66 (d, $^3J_{H,H}$ = 8.55 Hz, 1H, CHar), 7.16 (d, $^3J_{H,H}$ = 8.55 Hz, 1H, CHar), 4.16-4.05 (m, 4H, 2 x CH_2), 1.36 (t, $^3J_{H,H}$ = 6.35 Hz, 3H, CH_3) ^{13}C-NMR (125 MHz, DMSO-d6): δ [ppm] =164.1 (1C, CO), 154.2 (1C, CHole), 152.8 (1C, Cq), 148.2 (1C, Cq), 126.6 (1C, CHar), 123.7 (1C, Cq), 117.9 (1C, Cq), 113.8 (1C, CHar), 112.7 (1C, CHar), 99.3 (1C, CN), 64.1 (1C, CH_2), 63.8 (1C, CH_2), 14.4 (1C, CH_3).

(2*E*)-2-cyano-4-(4-decyloxy)prop-2-enamide

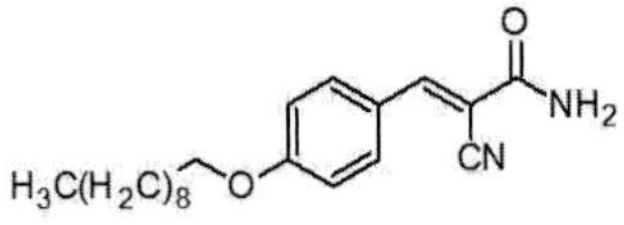

White crystals, 89% yield, Melting Point = 106.4 °C UV-Vis (MeOH: H_2O, 8:2,v/v):λ_{max} = 340 nm ^{1}H-NMR (500 MHz, DMSO-d6): δ [ppm] = 8.10 (s, 1H, CHole), 7.94 (d, $^3J_{H,H}$ = 8.55 Hz, 2H, 2 x CHar), 7.80 (s, br, 1H, NH), 7.67 (s, br, 1H, NH), 7.12 (d, $^3J_{H,H}$ =

8.55 Hz, 2H, 2 x CHar), 4.06 (t, $^3J_{H,H}$ = 6.35 Hz, 2H, CH_2), 1.75-1.69 (m, 2H, CH_2), 1.44-1.43 (m, 2H, CH_2), 1.30-1.18 (m, 12H, 6 x CH_2), 0.86 (t, $^3J_{H,H}$ = 6.35 Hz, 3H, CH_3) ^{13}C-NMR (125 MHz, DMSO-d6): δ [ppm] =163.2 (1C, CO), 161.9 (1C, Cq), 150.0 (1C, CHole), 132.6 (2C, 2 x CHar), 124.0 (1C, Cq), 116.8 (1C, Cq), 115.2 (2C, 2 x CHar), 102.6 (1C, CN), 68.0 (1C, CH_2), 31.3 (1C, CH_2), 29.0 (1C, CH_2), 28.8 (1C, CH_2), 28.6 (1C, CH_2), 28.4 (1C, CH_2), 25.3 (1C, CH_2), 22.0 (1C, CH_2), 22.5 (1C, CH_2), 13.8 (1C, CH_3).

(2*E*)-2-cyano-3-(3-ethoxy-4-propoxyphenyl)prop-2-enoic acid

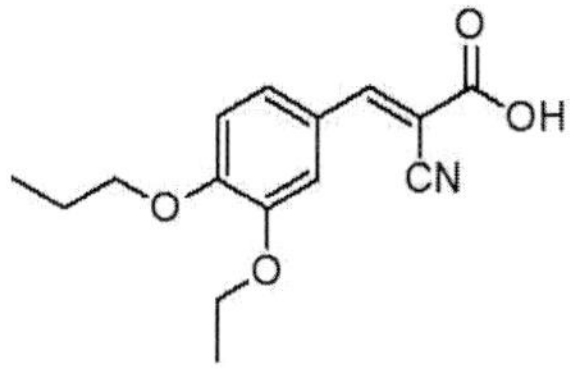

Pale yellow crystals, 87% yield, Melting Point = 157.5 °C UV-Vis (MeOH: H_2O, 8:2,v/v):λ_{max} = 341 nm ^{1}H-NMR (500 MHz, DMSO-d6): δ [ppm] = 8.21 (s, 1H, CHole), 7.75 (s, 1H, CHar), 7.65 (d, $^3J_{H,H}$ = 8.30 Hz, 1H, CHar), 7.16 (d, $^3J_{H,H}$ = 8.30 Hz, 1H, CHar), 4.08-4.02 (m, 4H, 2 x CH_2), 1.78-1.73 (m, 2H, CH_2), 1.35 (t, $^3J_{H,H}$ = 7.00 Hz, 3H, CH_3), 0.98 (t, $^3J_{H,H}$ = 7.30 Hz, 3H, CH_3) ^{13}C-NMR (125 MHz, DMSO-d6): δ [ppm] =163.1 (1C, CO), 152.4 (1C, Cq), 150.4 (1C, CHar), 134.8 (1C, Cq), 125.4 (1C, CHar), 114.1 (1C, CHar), 113.1 (1C, CHar), 102.9 (1C, CN), 69.8 (1C, CH_2), 63.8 (1C, CH_2), 21.8 (1C, CH_2), 14.6 (1C, CH_3), 10.6 (1C, CH_3).

(2*E*)-3-[4-(benzyloxy)-3-methoxyphenyl]-2-cyanoprop-2-enoic acid

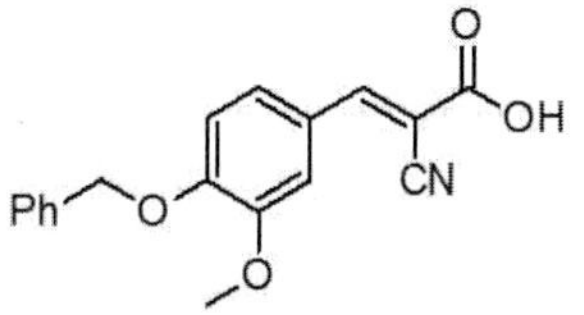

Pale yellow crystals, 83% yield, Melting Point = 206.7 °C UV-Vis (MeOH: H_2O, 8:2,v/v):λ_{max} = 340 nm ^{1}H-NMR (500 MHz, DMSO-d6): δ [ppm] = 13.72 (s, br 1H, COOH), 8.29 (s, 1H, CHole), 7.79 (s, 1H, CHar), 7.68 (d, $^3J_{H,H}$ = 8.30 Hz, 1H, CHar), 7.47 (d, $^3J_{H,H}$ = 7.25 Hz, 2H, 2 x CHar), 7.41 (t, $^3J_{H,H}$ = 7.25 Hz, 2H, 2 x CHar), 7.35 (t, $^3J_{H,H}$ = 7.15 Hz, 1H, CHar), 7.26 (d, $^3J_{H,H}$ = 8.55 Hz, 1H, CHar), 5.22 (s, 2H, CH_2), 3.86 (s, 3H, CH_3) ^{13}C-NMR (125 MHz, DMSO-d6): δ [ppm] =165.5 (1C, CO), 144.0 (1C, CHar), 141.4 (1C, CHar), 134.2 (1C, Cq), 133.9 (1C, Cq), 133.2 (1C, Cq), 133.1 (1C, Cq), 128.7 (2C, 2 x CHar), 128.5 (C, CHar), 128.1 (2C, 2 x CHar), 102.9 (1C, CN), 21.6 (1C, CH_2), 20.0 (1C, CH_3).

(2*E*)-2-cyano-3-[4-(trifluoromethoxy)phenyl]prop-2-enamide

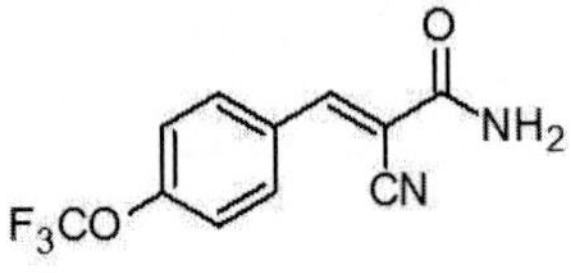

White crystals, 67% yield, Melting Point = 117.5 °C UV-Vis (MeOH: H_2O, 8:2,v/v):λ_{max} = 298 nm ^{1}H-NMR (500 MHz, DMSO-d6): δ [ppm] = 8.22 (s, 1H, CHole), 8.05 (d, $^3J_{H,H}$ = 8.55 Hz, 2H, 2 x CHar), 7.97 (s, br, 1H, NH), 7.83 (s, br, 1H, NH), 7.58 (d, $^3J_{H,H}$ = 8.55 Hz, 2H, 2 x CHar) ^{13}C-NMR (125 MHz, DMSO-d6): δ [ppm] = 162.9 (1C, CO), 152.9 (1C, CHole), 149.5 (1C, Cq), 132.6 (2C, 2 x CHar), 131.4 (1C, Cq), 121.9 (2C, 2 x CHar), 120.8 (1C, Cq), 116.5 (q, , $^1J_{C,F}$ = 271 Hz, 1C, CF_3), 108.3 (1C, CN).

(2*E*)-N'-acetyl-2-cyano-3-[4-(pentyloxy)phenyl]prop-2-enehydrazide

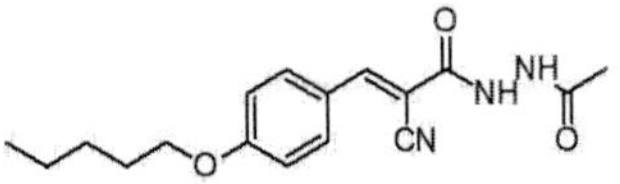

Pale yellow crystals, 91% yield, Melting Point = 158.7 °C UV-Vis (MeOH: H_2O, 8:2,v/v):λ_{max} = 362 nm ^{1}H-NMR (500 MHz, DMSO-d6): δ [ppm] = 10.25 (s, 1H, NH) 9.94 (s, 1H, NH), 8.10 (s, 1H, CHole), 7.68 (s, 1H, CHar), 7.56 (d, $^3J_{H,H}$ = 8.30 Hz, 1H, CHar), 7.16 (d, $^3J_{H,H}$ = 8.30 Hz, 1H, CHar), 4.09-4.05 (m, 4H, 2 x CH_2), 1.91 (s, 2H, CH_2), 1.75 (t, 3JH,H = 6.35 Hz, 2H, CH_2), 1.43-1.32 (m, 6H, 3 x CH_2), 0.93 (t, $^3J_{H,H}$ = 6.35 Hz, 3H, CH_3) ^{13}C-NMR (125 MHz, DMSO-d6): δ [ppm] = 168.6 (1C, CO), 153.3 (1C, Cq), 152.0 (1C, CHole), 148.7 (1C, CHar), 118.0 (1C, Cq), 113.7 (1C, CHar), 102.3 (1C, CN), 69.1 (1C, CH_2), 64.4 (1C, CH_2), 28.5 (2C, 2 x CH_2), 28.0 (2C, 2 x CH_2), 22.6 (1C, CH_2), 21.3 (1C, CH_2), 15.2 (1C, CH_3), 14.2 (1C, CH_3).

(2*E*)-2-cyano-3-[4-(2-methylpropoxy)phenyl]prop-2-enoic acid

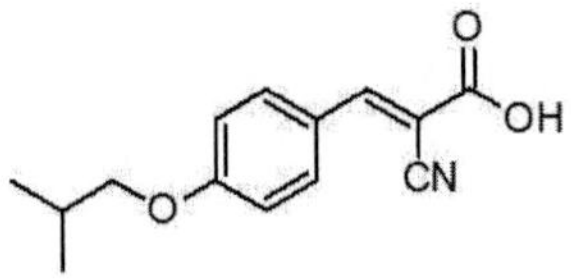

Pale yellow crystals, 78% yield, Melting Point = 168.6 °C UV-Vis (MeOH: H_2O, 8:2,v/v):λ_{max} = 342 nm ^{1}H-NMR (500 MHz, DMSO-d6): δ [ppm] = 13.78 (s, br 1H, COOH), 8.23 (s, 1H, CHole), 7.76 (s, 1H, CHar), 7.65 (d, $^3J_{H,H}$ = 8.30 Hz, 1H, CHar), 7.16 (d, $^3J_{H,H}$ = 7.25 Hz, 1H, CHar), 4.07 (q, $^3J_{H,H}$ = 7.00 Hz, 2H, CH_2), 3.86 (d, $^3J_{H,H}$ = 76.55 Hz, 2H, CH_2), 2.09-

2.03 (m, 1H, CH), 1.35 (t, $^{3}J_{H,H}$ = 6.90 Hz, 3H, CH_3), 0.99 (s, 3H, CH_3), 0.98 (s, 3H, CH_3) ^{13}C-NMR (125 MHz, DMSO-d6): δ [ppm] = 163.4 (1C, CO), 154.3 (1C, CHar), 153.0 (1C, Cq), 147.8 (1C, Cq), 126.7 (1C, CHar), 123.9 (1C, Cq), 116.8 (1C, CHar), 112.8 (1C, CHar), 99.4 (1C, CN), 74.5 (1C, CH_2), 74.1 (1C, CH_2), 27.8 (1C, CH), 18.7 (2C, 2 x CH_3), 14.6 (1C, CH_3).

(2*E*)-2-cyano-3-[4-(pentyloxy)phenyl]prop-2-enamide

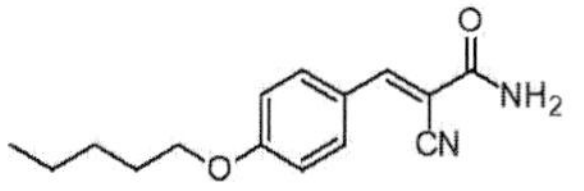

White crystals, 82% yield, Melting Point = 135.8 °C UV-Vis (MeOH: H_2O, 8:2,v/v):λ_{max} = 340 nm ^{1}H-NMR (500 MHz, DMSO-d6): δ [ppm] = 8.10 (s, 1H, CHole), 7.94 (d, $^{3}J_{H,H}$ = 8.30 Hz, 2H, 2 x CHar), 7.77 (s, br, 1H, NH), 7.65 (s, br, 1H, NH), 7.11 (d, $^{3}J_{H,H}$ = 8.30 Hz, 2H, 2 x CHar), 4.07 (t, , $^{3}J_{H,H}$ = 6.35 Hz, 2H, CH_2), 1.76-1.72 (m, 2H, CH_2), 1.43-1.32 (m, 4H, 2 x CH_2), 0.90 (t, $^{3}J_{H,H}$ = 7.10 Hz, 3H, CH_3) ^{13}C-NMR (125 MHz, DMSO-d6): δ [ppm] = 162.8 (1C, CO), 162.1 (1C, Cq), 150.0 (1C, CHole), 132.6 (2C, 2 x CHar), 124.2 (1C, Cq), 117.0 (1C, Cq), 115.3 (2C, 2 x CHar), 102.3 (1C, CN), 67.8 (1C, CH_2), 28.4 (1C, CH_2), 27.5 (1C, CH_2), 21.8 (1C, CH_2), 14.2 (1C, CH_3).

(2*E*)-2-cyano-3-[3-ethoxy-4-(pentyloxy)phenyl]prop-2-enamide

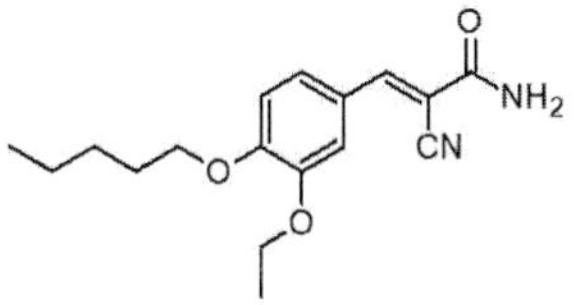

White crystals, 86% yield, Melting Point = 137.2 °C UV-Vis (MeOH: H_2O, 8:2,v/v):λ_{max} = 357 nm ^{1}H-NMR (500 MHz, DMSO-d6): δ [ppm] = 8.09 (s, 1H, CHole), 7.93 (s, br, 1H, NH), 7.67 (s, 2H, CHar, NH), 7.53 (d, $^{3}J_{H,H}$ = 8.30 Hz, 1H, CHar), 7.14 (d, $^{3}J_{H,H}$ = 7.25 Hz, 1H, CHar), 4.07-4.03 (m, 4H, 2 x CH_2), 1.77-1-70 (m, 2H, CH_2), 1.40-1.32 (m, 7H, 2 x CH_2, CH_3), 0.89 ((t, $^{3}J_{H,H}$ = 6.65 Hz, 3H, CH_3) ^{13}C-NMR (125 MHz, DMSO-d6): δ [ppm] = 162.9 (1C, CO), 152.0 (1C, CHar), 150.0 (1C, Cq), 148.0 (1C, Cq), 125.8 (1C, CHar), 124.2 (1C, Cq), 118.6 (1C, CHar), 116.8 (1C, CHar), 102.5(1C, CN), 68.1 (1C, CH_2), 63.9 (1C, CH_2), 28.2 (1C, CH), 27.4 (1C, CH_2), 21.8 (1C, CH_3), 14.6 (1C, CH_3), 13.9 (1C, CH_3).

(2*E*)-2-cyano-3-[4-(2-methylpropoxy)phenyl]prop-2-enamide

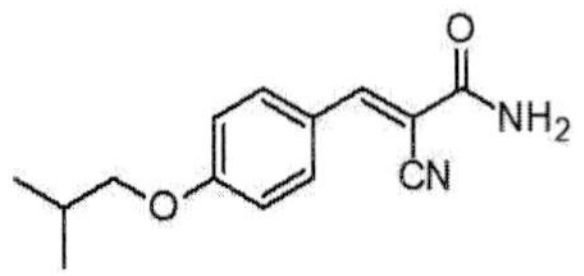

White crystals, 89% yield, Melting Point = 130.4 °C UV-Vis (MeOH: H_2O, 8:2,v/v):λ_{max} = 357 nm ^{1}H-NMR (500 MHz, DMSO-d6): δ [ppm] = 8.09 (s, 1H, CHole), 7.79 (s, 1H, NH), 7.68 (s, 2H, CHar, NH), 7.52 (d, $^3J_{H,H}$ = 7.25 Hz, 1H, CHar), 7.14 (d, $^3J_{H,H}$ = 7.25 Hz, 1H, CHar), 4.07 (q, $^3J_{H,H}$ = 7.00 Hz, 2H, CH_2), 3.86 (d, $^3J_{H,H}$ = 76.55 Hz, 2H, CH_2), 2.09-2.03 (m, 1H, CH), 1.35 (t, $^3J_{H,H}$ = 6.90 Hz, 3H, CH_3), 0.99 (s, 3H, CH3), 0.98 (s, 3H, CH_3) ^{13}C-NMR (125 MHz, DMSO-d6): δ [ppm] = 163.1 (1C, CO), 152.3 (1C, CHar), 150.0 (1C, Cq), 147.8 (1C, Cq), 125.8 (1C, CHar), 124.5 (1C, Cq), 116.8 (1C, CHar), 113.8 (1C, CHar), 112.9 (1C, CHar), 102.5 (1C, CN), 74.2 (1C, CH_2), 64.1 (1C, CH_2), 27.8 (1C, CH), 18.7 (2C, 2 x CH_3), 14.6 (1C, CH_3).

(2*E*)-2-cyano-3-[4-(prop-2-yn-1-yloxy)phenyl]prop-2-enamide

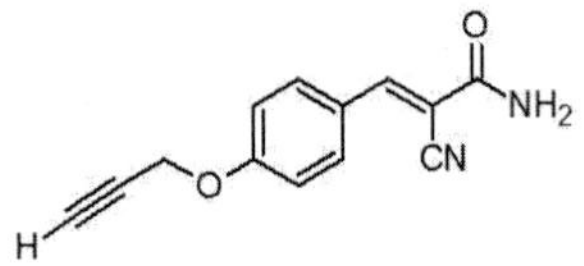

White crystals, 89% yield, Melting Point = 169.6 °C UV-Vis (MeOH: H_2O, 8:2,v/v):λ_{max} = 333 nm ^{1}H-NMR (500 MHz, DMSO-d6): δ [ppm] = 8.12 (s, 1H, CHole), 7.96 (d, $^3J_{H,H}$ = 8.55 Hz, 2H, 2 x CHar), 7.82 (s, br, 1H, NH), 7.69 (s, br, 1H, NH), 7.18 (d, $^3J_{H,H}$ = 8.55 Hz, 2H, 2 x CHar), 4.92 (s, 2H, CH_2), 3.65 (s, 1H, CH) ^{13}C-NMR (125 MHz, DMSO-d6): δ [ppm] = 162.9 (1C, CO), 160.6 (1C, Cq), 149.9 (1C, CHole), 132.6 (2C, 2 x CHar), 125.0 (1C, Cq), 116.8 (1C, Cq), 115.6 (2C, 2 x CHar), 103.3 (1C, CN), 78.7 (1C, CH), 78.5 (1C, Cq), 55.5 (1C, CH_2).

(2*E*)-3-[4-(benzyloxy)-3-methoxyphenyl]-2-cyanoprop-2-enamide - methane (1:1)

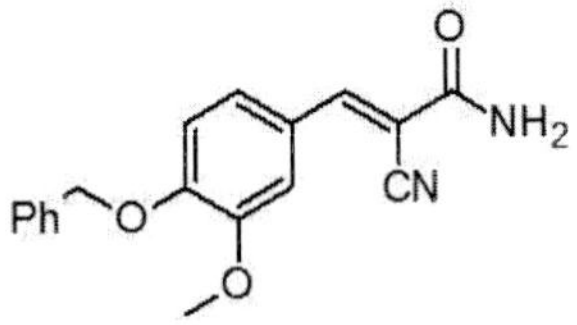

White crystals, 86% yield, Melting Point = 180.0 °C UV-Vis (MeOH: H_2O, 8:2,v/v):λ_{max} = 355 nm ^{1}H-NMR (500 MHz, DMSO-d6): δ [ppm] = 8.11 (s, 1H, CHole), 7.80 (s, br, 1H, NH), 7.68

(s, 1H, CHar), 7.67 (s, br, 1H, NH), 7.55 (d, $^3J_{H,H}$ = 8.30 Hz, 1H, CHar), 7.47 (d, $^3J_{H,H}$ = 7.25 Hz, 2H, 2 x CHar), 7.41 (d, $^3J_{H,H}$ = 7.25 Hz, 2H, 2 x CHar), 7.35 (t, $^3J_{H,H}$ = 7.15 Hz, 1H, CHar), 7.26 (d, $^3J_{H,H}$ = 8.55 Hz, 1H, CHar), 5.24 (s, 2H, CH_2), 3.84 (s, 3H, CH_3) ^{13}C-NMR (125 MHz, DMSO-d6): δ [ppm] = 165.3 (1C, CO), 144.2 (1C, CHar), 141.4 (1C, CHar), 136.2 (1C, Cq), 133.9 (1C, Cq), 133.2 (1C, Cq), 133.1 (1C, Cq), 128.7 (2C, 2 x CHar), 128.5 (1C, CHar), 128.1 (2C, 2 x CHar), 102.9 (1C, CN), 21.6 (1C, CH_2), 20.0 (1C, CH_3).

6.1.2.3 Diene Derivatives

(2*E*,4*E*)-2-cyano-5-phenylpenta-2,4-dienamide

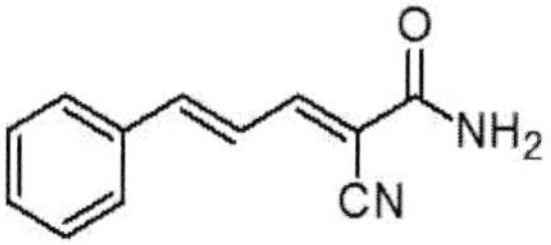

O NH2 CN

Pale yellow crystals, 72% yield, Melting Point = 135.7 °C UV-Vis (MeOH: H_2O, 8:2,v/v):λ_{max} = 340 nm ^{1}H-NMR (500 MHz, DMSO-d6): δ [ppm] = 7.97 (d, $^3J_{H,H}$ = 11.30 Hz, 1H, CHole), 7.80 (s, br, 1H, NH), 7.69 (d, $^3J_{H,H}$ = 3.90 Hz, 2H, CHole, CHar), 7.64 (s, br, 1H, NH), 7.47-7.43 (m, 4H, 4 x CHar), 7.17 (q, $^3J_{H,H}$ = 8.90 Hz, 1H, CHole) ^{13}C-NMR (125 MHz, DMSO-d6): δ [ppm] = 162.9 (1C, CO), 151.9 (1C, CHar), 147.2 (1C, CHar), 135.4 (1C, Cq), 131.1 (1C, CHar), 129.5 (2C, 2 x CHar), 127.8 (2C, 2 x CHar), 123.6 (1C, CHar), 116.0 (1C, Cq), 108.3 (1C, CN).

(2*E*,4*E*)-2-cyano-5-[4-(dimethylamino)phenyl]penta-2,4-dienoic acid

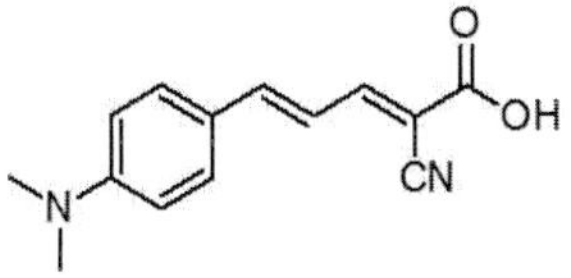

O OH CN N

Black crystals, 76% yield, Melting Point = 163.1 °C UV-Vis (MeOH: H_2O, 8:2,v/v):λ_{max} = 427 nm ^{1}H-NMR (500 MHz, DMSO-d6): δ [ppm] = 7.90 (d, $^3J_{H,H}$ = 11.80 Hz, 1H, CHole), 7.50 (d, $^3J_{H,H}$ = 8.75 Hz, 2H, 2 x CHar), 7.41 (d, $^3J_{H,H}$ = 14.80 Hz, 1H, CHole), 7.17 (q, $^3J_{H,H}$ = 8.90 Hz, 1H, CHole), 6.75 (d, $^3J_{H,H}$ = 8.75 Hz, 2H, 2 x CHar), 3.01 (s, 6H, 2 x CH_3)

^{13}C-NMR (125 MHz, DMSO-d6): δ [ppm] = 165.0 (1C, CO), 164.1 (1C, Cq), 154.1 (1C, CHole), 152.0 (1C, Cq), 148.8 (1C, CHole), 130.5 (2C, 2 x CHar), 122.2 (1C, Cq), 117.4 (1C, CHole), 116.8 (1C, Cq), 116.5 (1C, Cq), 111.9 (2C, 2 x CHar), 102.2(1C, CN), 25.9 (2C,2 x CH_3).

(2*E*)-2-cyano-3-4-[(*E*)-2-phenylethenyl]phenylprop-2-enamide

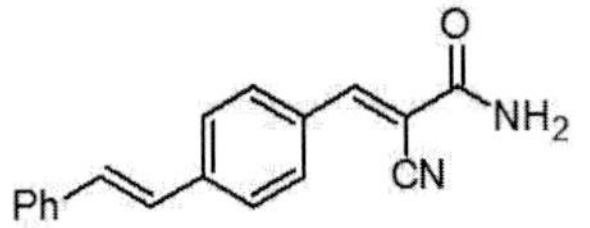

Pale yellow crystals, 87% yield, Melting Point = 209.5 °C UV-Vis (MeOH: H_2O, 8:2,v/v):λ_{max} = 362 nm ^{1}H-NMR (500 MHz, DMSO-d6): δ [ppm] = 8.16 (s, 1H, CHole), 7.97 (d, $^3J_{H,H}$ = 8.10 Hz, 2H, 2 x CHar), 7.91 (s, br, 1H, NH), 7.80 (d, $^3J_{H,H}$ = 8.10 Hz, 2H, 2 x CHar), 7.77 (s, br, 1H, NH), 7.66 (d, $^3J_{H,H}$ = 7.25 Hz, 2H, 2 x CHar), 7.49-7.30 (m, 5H, 3 x CHar, 2 x CHole) ^{13}C-NMR (125 MHz, DMSO-d6): δ [ppm] = 162.7 (1C, CO), 149.8 (1C, CHole), 145.7 (1C, Cq), 141.0 (1C, Cq), 136.5 (1C, Cq), 131.5 (1C, CHole), 130.5 (1C, Cq), 130.4 (2C, 2 x CHar), 129.3 (2C, 2 x CHar), 127.2 (1C, CHole), 127.0 (2C, 2 x CHar), 126.8 (2C, 2 x CHar), 126.7 (1C, Cq), 116.8 (1C, Cq), 116.5 (1C, Cq), 105.3 (1C, CN).

(2*E*,4*E*)-2-cyano-5-(pyridin-3-yl)penta-2,4-dienamide

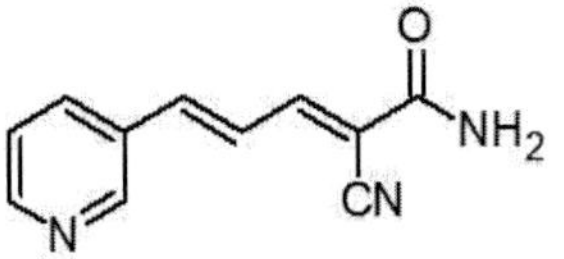

Pale yellow crystals, 74% yield, Melting Point = 246.1 °C UV-Vis (MeOH: H_2O, 8:2,v/v):λ_{max} = 327 nm ^{1}H-NMR (500 MHz, DMSO-d6): δ [ppm] = 8.83 (s, 1H, CHole), 8.59 (d, $^3J_{H,H}$ = 8.10 Hz, 1H, CHar), 8.20 (d, $^3J_{H,H}$ = 8.10 Hz, 1H, CHar), 7.98 (d, $^3J_{H,H}$ = 8.10 Hz, 1H, CHole), 7.88 (s, br, 1H, NH), 7.70 (s, br, 1H, NH), 7-49-7.51 (m, 2H, 2 x CHar), 7.29 (q, $^3J_{H,H}$ = 8.90 Hz, 1H, CHole) ^{13}C-NMR (125 MHz, DMSO-d6): δ [ppm] = 162.3 (1C, CO), 150.8 (1C, CHole), 150.6 (1C, CHole), 149.8 (1C, CHar), 143.0 (1C, CHar), 134.3 (1C, CHar), 130.7 (1C, Cq), 124.7 (1C, CHole), 123.9 (1C, CHar), 115.1 (1C, Cq), 108.7 (1C, CN).

2-[(2*E*)-3-(4-methoxyphenyl)prop-2-en-1-ylidene]propanediamide

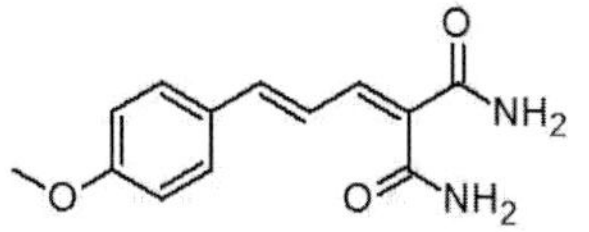

Yellow crystals, 59% yield, Melting Point = 232.4 °C UV-Vis (MeOH: H_2O, 8:2,v/v):λ_{max} = 343 nm ^{1}H-NMR (500 MHz, DMSO-d6): δ [ppm] = 7.91 (s, br, 1H, NH), 7.50 (s, br, 1H, NH), 7.47 (d, $^3J_{H,H}$ = 8.10 Hz, 2H, 2 x CHar), 7.41 (s, br, 1H, NH), 7.24 (s, br, 1H, NH), 7.18-7.16 (m, 2H, 2 x CHole), 7.01 (q, $^3J_{H,H}$ = 8.90 Hz, 1H, CHole), 6.97 (d, $^3J_{H,H}$ = 8.10 Hz, 2H, 2 x CHar), 3.80 (s, 3H, CH_3) ^{13}C-NMR (125 MHz, DMSO-d6): δ [ppm] = 167.9 (1C, CO), 166.7 (1C, Cq), 160.0 (1C, Cq), 140.3 (1C, CHole), 138.7 (1C, CHole), 129.8 (1C, Cq), 128.7 (2C, 2 x CHar), 122.1 (1C, CHole), 114.4 (2C, 2 x CHar), 55.2 (1C, CH_3).

(2*E*)-2-cyano-3-4-[(*E*)-2-phenylethenyl]phenylprop-2-enoic acid

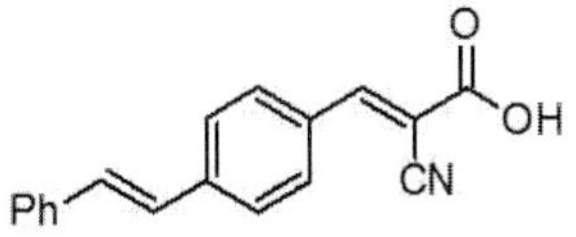

Bright yellow crystals, 77% yield, Melting Point = 216.3 °C UV-Vis (MeOH: H_2O, 8:2,v/v):λ_{max} = 349 nm ^{1}H-NMR (500 MHz, DMSO-d6): δ [ppm] = 8.16 (s, 1H, CHole), 7.97 (d, $^3J_{H,H}$ = 8.10 Hz, 2H, 2 x CHar), 7.91 (s, br, 1H, NH), 7.80 (d, $^3J_{H,H}$ = 8.10 Hz, 2H, 2 x CHar), 7.77 (s, br, 1H, NH), 7.66 (d, $^3J_{H,H}$ = 7.25 Hz, 2H, 2 x CHar), 7.49-7.30 (m, 5H, 3 x CHar, 2 x CHole) ^{13}C-NMR (125 MHz, DMSO-d6): δ [ppm] = 162.7 (1C, CO), 149.8 (1C, CHole), 145.7 (1C, Cq), 141.0 (1C, Cq), 136.5 (1C, Cq), 131.5 (1C, CHole), 130.5 (1C, Cq), 130.4 (2C, 2 x CHar), 129.3 (2C, 2 x CHar), 127.2 (1C, CHole), 127.0 (2C, 2 x CHar), 126.8 (2C, 2 x CHar), 126.7 (1C, Cq), 116.8 (1C, Cq), 116.5 (1C, Cq), 105.3 (1C, CN).

(2*E*,4*E*)-2-cyano-3-methyl-5-(thiophen-2-yl)penta-2,4-dienamide

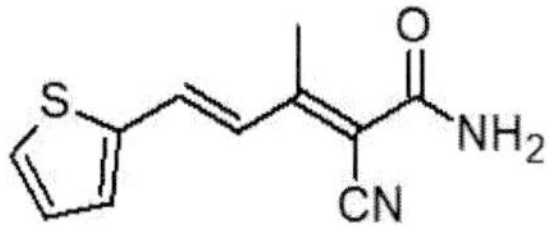

Yellow crystals, 65% yield, Melting Point = 217.3 °C UV-Vis (MeOH: H_2O, 8:2,v/v):λ_{max} = 359 nm ^{1}H-NMR (500 MHz, DMSO-d6): δ [ppm] = 7.97 (s, br, 1H, NH), 7.74 (s, br, 1H, NH), 7.67 (d, $^3J_{H,H}$ = 8.10 Hz, 1H, CHar), 7.51 (d, $^3J_{H,H}$ = 14.80 Hz, 1H, CHole), 7.40-7.37 (m, 2H, CHole, CHar), 7.15-7.13 (m, 1H, CHar), 2.29 (s, 3H, CH_3) ^{13}C-NMR (125 MHz, DMSO-d6): δ [ppm] = 162.7 (1C, CO), 155.2 (1C, Cq), 141.0 (1C,

Cq), 131.9 (1C, CHar), 130.9 (1C, CHar), 128.8 (1C, CHole), 128.5 (1C, CHole), 123.4 (1C, CHar), 117.5 (1C, Cq), 107.1 (1C, CN), 18.1 (1C, CH_3).

(2*E*)-2-cyano-5,5-diphenylpenta-2,4-dienamide

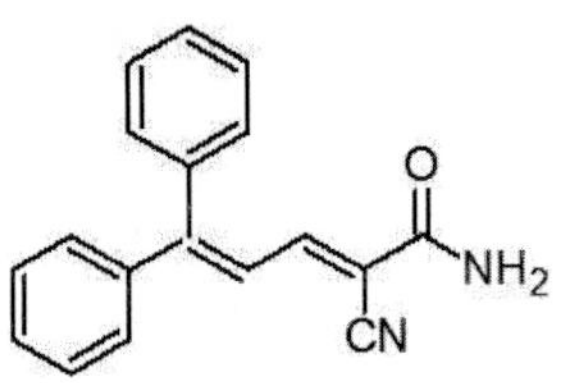

Pale yellow crystals, 76% yield, Melting Point = 241.1 °C UV-Vis (MeOH: H_2O, 8:2,v/v):λ_{max} = 354 nm ^{1}H-NMR (500 MHz, DMSO-d6): δ [ppm] = 7.81 (s, br, 1H, NH), 7.66-7.62 (m, 2H, CHole, NH), 7.53-7.23 (m, 10H, 10 x CHar), 7.06 (d, $^3J_{H,H}$ = 11.80 Hz, 1H, CHole) ^{13}C-NMR (125 MHz, DMSO-d6): δ [ppm] = 162.1 (1C, CO), 156.4 (1C, Cq), 148.2 (1C, CHole), 139.8 (1C, Cq), 137.3 (1C, Cq), 130.2 (2C, 2 x CHar), 130.0 (1C, CHole), 129.2 (1C, CHar), 128.5 (1C, CHar), 128.7 (2C, 2 x CHar), 128.6 (2C, 2 x CHar), 128.4 (2C, 2 x CHar), 121.8 (1C, CHar), 115.5 (1C, Cq), 108.7 (1C, CN).

(2*E*,4*E*)-2-cyano-5-(4-fluorophenyl)penta-2,4-dienamide

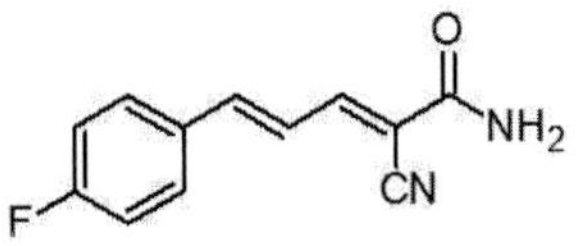

Pale yellow crystals, 77% yield, Melting Point = 187.2 °C UV-Vis (MeOH: H_2O, 8:2,v/v):λ_{max} = 339 nm ^{1}H-NMR (500 MHz, DMSO-d6): δ [ppm] = 7.96 (d, $^3J_{H,H}$ = 11.30 Hz, 1H, CHole), 7.84 (s,br, 1H, NH), 7.78 (t, $^3J_{H,H}$ = 6.50 Hz, 1H, CHar), 7.65 (s,br, 1H, NH), 7.44 (d, $^3J_{H,H}$ = 15.40 Hz, 1H, CHole), 7.29 (t, $^3J_{H,H}$ = 8.65 Hz, 1H, CHar), 7.16-7.10 (m, 1H, CHole) ^{13}C-NMR (125 MHz, DMSO-d6): δ [ppm] = 164.2 (1C, CO), 162.4 (1C, Cq), 151.1 (1C, CHole), 145.4 (1C, CHole), 131.4 (d, $^4J_{C,F}$ = 3.10 Hz, 1C, CHar), 130.5 (d, $^3J_{C,F}$ = 8.75 Hz, 2C, 2 x CHar), 122.8 (1C, CHar), 116.1 (d, $^2J_{C,F}$ = 21 Hz, 2C, 2 x CHar), 115.3 (1C, Cq), 107.6 (1C, CN).

(2*E*,4*E*)-2-cyano-5-(4-fluorophenyl)penta-2,4-dienoic acid

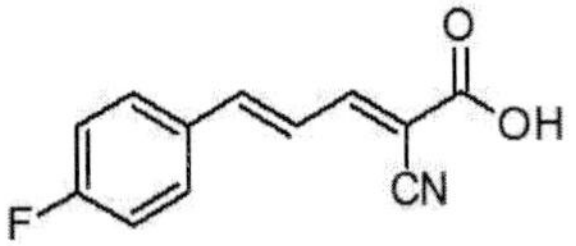

Yellow crystals, 59% yield, Melting Point = 241.1 °C UV-Vis (MeOH: H_2O, 8:2,v/v):λ_{max} = 337 nm ^{1}H-NMR (500 MHz, DMSO-d6): δ [ppm] = 8.08 (d,

$^3J_{H,H}$ = 11.30 Hz, 1H, CHole), 7.81- 7.77 (m, 1H, CHar), 7.64 (d, $^3J_{H,H}$ = 15.40 Hz, 1H, CHole), 7.29 (t, $^3J_{H,H}$ = 8.65 Hz, 1H, CHar), 7.20-7.14 (m, 1H, CHole) ^{13}C-NMR (125 MHz, DMSO-d6): δ [ppm] = 164.4 (1C, CO), 163.1 (1C, Cq), 162.4 (1C, Cq), 155.1 (1C, CHole), 147.6 (1C, CHole), 131.4 (d, $4J_{C,F}$ = 3.10 Hz, 2 x C, CHar), 130.5 (d, $^3J_{C,F}$ = 8.75 Hz, 2C, 2 x CHar), 122.6 (1C, CHar), 116.1 (d, $^2J_{C,F}$ = 21 Hz, 2C, 2 x CHar), 114.9 (1C, Cq), 104.5 (1C, CN).

(2E,4E)-2-cyano-5-[4-(dimethylamino)phenyl]penta-2,4-dienamide

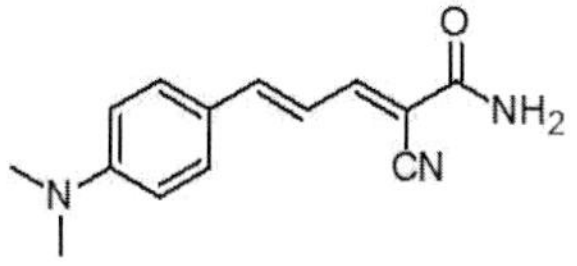

O NH$_2$ N CN

106 Red crystals, 92% yield, Melting Point = 241.1 °C UV-Vis (MeOH: H_2O, 8:2,v/v):λ_{max} = 461 nm ^{1}H-NMR (500 MHz, DMSO-d6): δ [ppm] = 7.90 (d, $^3J_{H,H}$ = 11.80 Hz, 1H, CHole), 7.61 (s, br, 1H, NH), 7.50 (d, $^3J_{H,H}$ = 8.75 Hz, 2H, 2 x CHar), 7.44 (s, br, 1H, NH), 7.31 (d, $^3J_{H,H}$ = 14.80 Hz, 1H, CHole), 6.98 (q, $^3J_{H,H}$ = 8.90 Hz, 1H, CHole), 6.75 (d, $^3J_{H,H}$ = 8.75 Hz, 2H, 2 x CHar), 3.01 (s, 6H, 2 x CH_3) ^{13}C-NMR (125 MHz, DMSO-d6): δ [ppm] = 163.8 (1C, CO), 152.9 (1C, CHole), 149.0 (1C, CHole), 130.9 (2C, 2 x CHar), 122.2 (1C, Cq), 117.8 (1C, CHole), 116.8 (1C, Cq), 116.5 (1C, Cq), 112.1 (2C, 2 x CHar), 103.2(1C, CN), 25.9 (2C,2 x CH_3).

(2E,4E)-2-cyano-5-(furan-2-yl)penta-2,4-dienoic acid

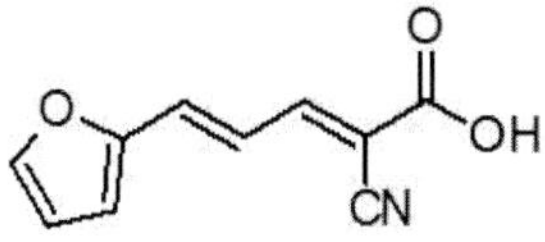

O O OH CN

Orange crystals, 89% yield, Melting Point = 246.1 °C UV-Vis (MeOH: H_2O, 8:2,v/v):λ_{max} = 358 nm ^{1}H-NMR (500 MHz, DMSO-d6): δ [ppm] = 8.08 (d, $^3J_{H,H}$ = 11.30 Hz, 1H, CHole), 7.96 (s, 1H, CHar), 7.50 (d, $^3J_{H,H}$ = 15.40 Hz, 1H, CHole), 7.04 (d, $^3J_{H,H}$ = 8.65 Hz, 1H, CHar), 6.93 (q, $^3J_{H,H}$ = 8.90 Hz, 1H, CHole), 6.71 (s, 1H, CHar) ^{13}C-NMR (125 MHz, DMSO-d6): δ [ppm] = 163.1 (1C, CO), 154.5 (1C, CHole), 150.7 (1C, Cq), 147.0 (1C, CHole), 134.4 (1C, CHar), 120.5 (1C, CHar), 117.5 (1C, CHole), 114.9 (1C, Cq), 113.3 (1C, CHar), 103.6 (1C, CN).

(2*E*,4*E*)-2-cyano-5-(4-hydroxy-3-methoxyphenyl)penta-2,4-dienoic acid

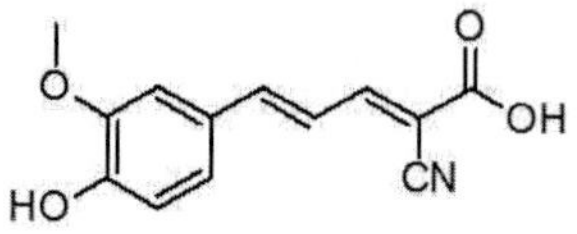

Yellow crystals, 89% yield, Melting Point = 230.3 °C UV-Vis (MeOH: H_2O, 8:2,v/v):λ_{max} = 368 nm ^{1}H-NMR (500 MHz, DMSO-d6): δ [ppm] = 9.90 (s, 1H, OH), 8.03 (d, $^3J_{H,H}$ = 11.30 Hz, 1H, CHole), 7.54 (d, $^3J_{H,H}$ = 15.40 Hz, 1H, CHole), 7.24 (s, 1H, CHar), 7.16 (d, $^3J_{H,H}$ = 3.20 Hz, 1H, CHar), 7.05 (q, $^3J_{H,H}$ = 8.90 Hz, 1H, CHole), 6.85 (d, $^3J_{H,H}$ = 8.20 Hz, 1H, CHar), 3.83 (s, 3H, CH_3) ^{13}C-NMR (125 MHz, DMSO-d6): δ [ppm] = 164.1 (1C, CO), 156.0 (1C, CHole), 150.3 (1C, Cole), 148.0 (1C, Cq), 126.4 (1C, Cq), 124.2 (1C, CHar), 119.5 (1C, CHole), 115.9 (1C, CHar), 111.7 (1C, CHar), 101.6 (1C, CN), 55.3 (1C, CH_3).

(2*E*,4*E*)-2-cyano-5-(furan-2-yl)penta-2,4-dienamide

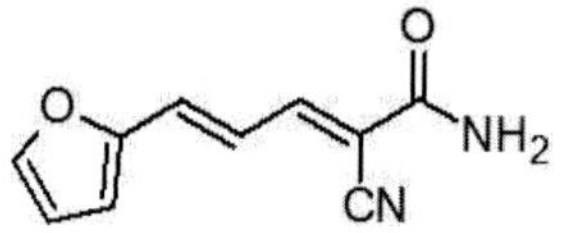

Yellow crystals, 73% yield, Melting Point = 241.1 °C UV-Vis (MeOH: H_2O, 8:2,v/v):λ_{max} = 371 nm ^{1}H-NMR (500 MHz, DMSO-d6): δ [ppm] = 7.95 (d, $^3J_{H,H}$ = 11.30 Hz, 1H, CHole), 7.92 (s, 1H, CHar), 7.79 (s,br, 1H, NH), 7.62 (s,br, 1H, NH), 7.30 (d, $^3J_{H,H}$ = 15.40 Hz, 1H, CHole), 6.98 (d, $^3J_{H,H}$ = 8.65 Hz, 1H, CHar), 6.93 (q, $^3J_{H,H}$ = 8.90 Hz, 1H, CHole), 6.68 (s, 1H, CHar) ^{13}C-NMR (125 MHz, DMSO-d6): δ [ppm] = 162.6 (1C, CO), 150.8 (1C, CHole), 146.8 (1C, CHole), 132.6 (1C, CHar), 120.5 (1C, CHar), 116.5 (1C, CHole), 115.5 (1C, Cq), 113.3 (1C, CHar), 107.0 (1C, CN).

(2*E*,4*E*)-2-cyano-5-(4-hydroxy-3,5-dimethoxyphenyl)penta-2,4-dienamide

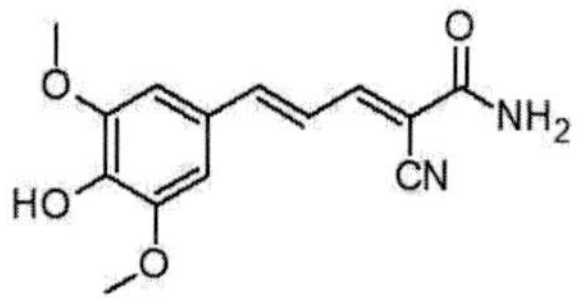

Yellow crystals, 87% yield, Melting Point = 213.4 °C UV-Vis (MeOH: H_2O, 8:2,v/v):λ_{max} = 395 nm ^{1}H-NMR (500 MHz, DMSO-d6): δ [ppm] = 7.90 (d, $^3J_{H,H}$ = 11.30 Hz, 1H, CHole), 7.72 (s,br, 1H, NH), 7.55 (s,br, 1H, NH), 7.30 (d, $^3J_{H,H}$ = 15.40 Hz, 1H, CHole), 7.06 (q, $^3J_{H,H}$ = 8.90 Hz, 1H, CHole), 6.98 (s, 2H, 2 x CHar) ^{13}C-NMR (125 MHz, DMSO-d6): δ [ppm] = 162.9 (1C, CO), 152.0 (1C, CHole), 148.3 (1C, CHole), 138.8 (1C, Cq), 125.5 (1C, Cq),

120.5 (1C, CHar), 115.8 (1C, Cq), 106.3 (2C, 2 x CHar), 105.0 (1C, CN), 56.2 (2C, 2 x CH_3).

(2*E*,4*E*)-2-cyano-5-(2-nitrophenyl)penta-2,4-dienamide

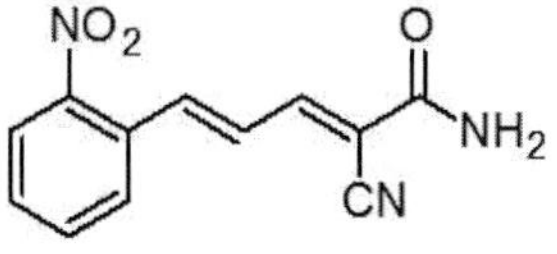

Pale yellow crystals, 57% yield, Melting Point = 201.8 °C UV-Vis (MeOH: H_2O, 8:2,v/v):λ_{max} = 311 nm ^{1}H-NMR (500 MHz, DMSO-d6): δ [ppm] = 8.09-8.01 (m, 3H, 3 x CHar), 7.91 (s,br, 1H, NH), 7.81 (d, $^3J_{H,H}$ = 7.55 Hz, 1H, CHole), 7.72-7.67 (m, 3H, CHar, CHole, NH), 7.15 (q, $^3J_{H,H}$ = 8.90 Hz, 1H, CHole) ^{13}C-NMR (125 MHz, DMSO-d6): δ [ppm] = 162.7 (1C, CO), 150.7 (1C, CHole), 148.7 (1C, Cq), 141.0 (1C, CHole), 134.4 (1C, CHar), 131.1 (1C, Cq), 129.4 (1C, CHar), 127.7 (1C, CHar), 125.4 (1C, CHole), 115.5 (1C, Cq), 111.2 (1C, CN).

(2*E*,4*E*)-2-cyano-5-(4-methoxyphenyl)penta-2,4-dienoic acid

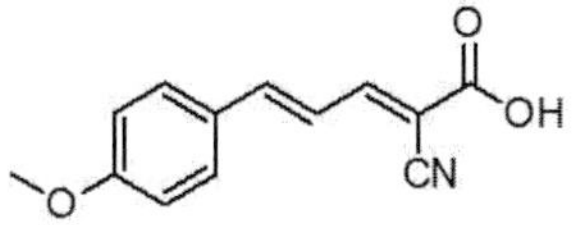

Bright yellow crystals, 82% yield, Melting Point = 233.8 °C UV-Vis (MeOH: H_2O, 8:2,v/v):λ_{max} = 356 nm ^{1}H-NMR (500 MHz, DMSO-d6): δ [ppm] = 8.05 (d, $^3J_{H,H}$ = 11.30 Hz, 1H, CHole), 7.67 (d, $^3J_{H,H}$ = 8.65 Hz, 2H, 2 x CHar), 7.60 (d, $^3J_{H,H}$ = 15.10 Hz, 1H, CHole), 7.08-7.02 (m, 3H, CHole, 2 x CHar), 3.83 (s, 3H, CH_3) ^{13}C-NMR (125 MHz, DMSO-d6): δ [ppm] = 163.9 (1C, CO), 162.6 (1C, Cq), 156.8 (1C, CHole), 149.8 (1C, CHole), 132.1 (1C, CHole), 130.9 (2C, 2 x CHar), 128.1 (1C, Cq), 121.2 (1C, CHole), 115.3 (2C, 2 x CHar), 103.3 (1C, CN).

(2*E*,4*E*)-2-cyano-5-(4-nitrophenyl)penta-2,4-dienamide

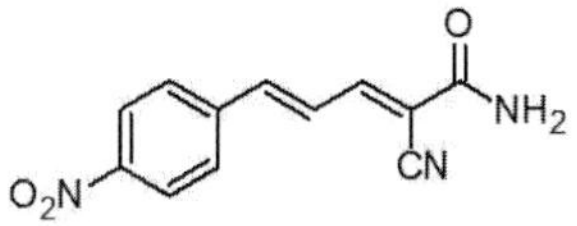

Yellow crystals, 82% yield, Melting Point = 256.9 °C UV-Vis (MeOH: H_2O, 8:2,v/v): λ_{max} = 345 nm ^{1}H-NMR (500 MHz, DMSO-d6): δ [ppm] = 8.27 (d, $^3J_{H,H}$ = 8.50 Hz, 1H, CHar), 8.01-7-97 (m, 3H, CHole, 2 x CHar), 7.93 (s,br, 1H, NH), 7.75 (s,br, 1H, NH), 7.55 (d, $^3J_{H,H}$ = 15.10 Hz, 1H, CHole), 7.35 (q, $^3J_{H,H}$ = 8.90 Hz, 1H, CHole) ^{13}C-NMR (125

MHz, DMSO-d6): δ [ppm] = 162.2 (1C, CO), 150.3 (1C, CHole), 147.7 (1C, Cq), 143.5 (1C, CHole), 141.0 (1C, Cq), 129.4 (2C, 2 x CHar), 126.9 (1C, CHole), 124.0 (2C, 2 x CHar), 115.0 (1C, Cq), 110.3 (1C, CN).

(2*E*)-2-cyano-5,5-diphenylpenta-2,4-dienoic acid

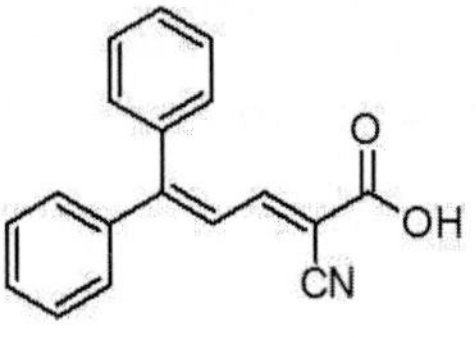

UV-Vis (MeOH: H_2O, 8:2,v/v):λ_{max} = 353 nm Pale yellow crystals, 76% yield, Melting Point = 222.51 °C ^{1}H-NMR (500 MHz, DMSO-d6): δ [ppm] = 7.66-7.38 (m, 9H, 8 x CHar, CHole), 7.66 (s, 1H, CHar), 7.11 (d, $^3J_{H,H}$ = 11.80 Hz, 1H, CHole) ^{13}C-NMR (125 MHz, DMSO-d6): δ [ppm] = 163.1 (1C, CO), 158.4 (1C, Cq), 151.3 (1C, CHole), 139.1 (1C, Cq), 137.3 (1C, Cq), 130.4 (2C, 2 x CHar), 130.0 (1C, CHole), 129.2 (1C, CHar), 128.5 (1C, CHar), 128.7 (2C,2 x CHar), 128.6 (2C,2 x CHar), 128.4 (2C,2 x CHar), 121.4 (1C, CHar), 115.0 (1C, Cq), 105.7 (1C, CN).

6.1.2.4 Pyrrole derivatives

(2*E*)-2-cyano-3-1-[4-(trifluoromethyl)phenyl]-1*H*-pyrrol-2-ylprop-2-enamide (64)

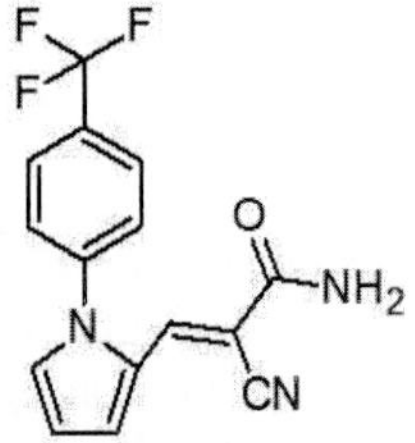

Pale yellow crystals, 89% yield, Melting Point = 178.7 °C UV-Vis (MeOH: H_2O, 8:2,v/v):λ_{max} = 365 nm MALDI-MS: $[M+H]^+$ = 306.33 Da ^{1}H-NMR (500 MHz, DMSO-d6): δ [ppm] = 7.97 (d, $^3J_{H,H}$ = 8.25, 2H, 2 x CHar), 7.64-7.54 (m, 7H, 2 x NH, CHole, 4 x CHar), 6.65-6.63 (m, 1H, CHar) ^{13}C-NMR (125 MHz, DMSO-d6): δ [ppm] = 160.0 (1C, CO), 137.5 (1C, CHole), 130.2 (1C, CHar), 127.3 (2C, 2 x CHar), 126.5 (1C, Cq), 122.5 (q, $^1J_{C,F}$ = 269 Hz, 1C, CF_3), 117.2 (1C, CHar), 106.3 (1C, CHar), 112.5 (1C, CHar), 98.3 (1C, CN).

(2*E*)-2-cyano-3-1-[4-(trifluoromethyl)phenyl]-1*H*-pyrrol-2-ylprop-2-enoic acid (65)

Yellow crystals, 78% yield, Melting Point = 209.1 °C UV-Vis (MeOH: H_2O, 8:2,v/v):λ_{max} = 350 nm MALDI-MS: $[M+H]^+$ = 306.17 Da ^{1}H-NMR (500 MHz, DMSO-d6): δ [ppm] = 7.97. (d, $^3J_{H,H}$ = 8.25, 2H, 2 x CHar), 7.67-7.54 (m, 5H, CHole, 4 x CHar), 6.61-6.58 (m, 1H, CHar) ^{13}C-NMR (125 MHz, DMSO-d6): δ [ppm] = 164.2 (1C, CO), 140.5 (1C, CHole), 131.2 (1C, CHar), 127.9 (2C, 2 x CHar), 126.7 (1C, Cq), 122.7 (q, $^1J_{C,F}$ = 269 Hz, 1C, CF_3), 118.2 (1C, CHar), 116.9 (1C, Cq), 113.0 (1C, CHar), 95.3 (1C, CN).

(2*E*)-2-cyano-3-[1-(4-methylphenyl)-1*H*-pyrrol-2-yl]prop-2-enamide (66)

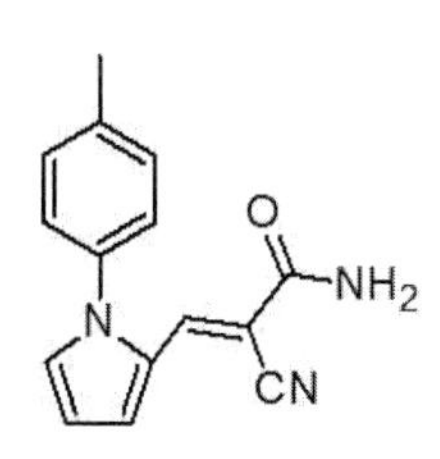

Pale yellow crystals, 66% yield, Melting Point = 193.8 °C UV-Vis (MeOH: H_2O, 8:2,v/v):λ_{max} = 368 nm ^{1}H-NMR (500 MHz, DMSO-d6): δ [ppm] = 7.70 (s, 1H, CHole), 7.54-7.28 (m, 9H, 6 x CHar, 2 x NH), 6.59-6.57 (m, 1H, CHar), 2.40 (1H, CH_3) ^{13}C-NMR (125 MHz, DMSO-d6): δ [ppm] = 162.9 (1C, CO), 138.4 (1C, CHole), 138.0 (1C, Cq), 134.7 (1C, Cq), 130.4 (1C, CHar), 130.0 (2C, 2 x CHar), 126.9 (1C, Cq), 126.5 (2C, 2 x CHar), 117.6 (1C, Cq), 116.4 (1C, CHar), 111.9 (1C, CHar), 97.1 (1C, CN), 20.5 (1C, CH_3).

(2*E*)-2-cyano-3-[1-(4-methylphenyl)-1*H*-pyrrol-2-yl]prop-2-enoic acid (67)

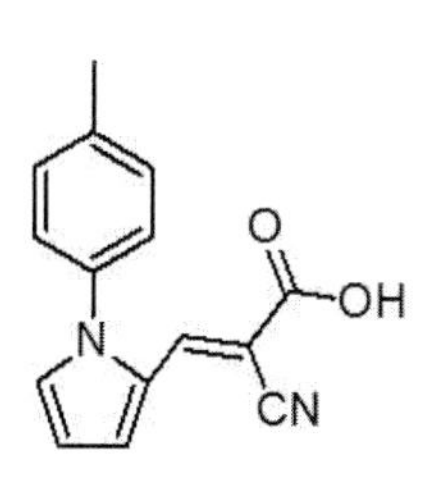

Yellow crystals, 77% yield, Melt-ing Point = 232.4 °C UV-Vis (MeOH: H_2O, 8:2,v/v):λ_{max} = 356 nm ^{1}H-NMR (500 MHz, DMSO-d6): δ [ppm] = 7.72 (s, 1H, CHole), 7.65-7.58 (m, 2H, 2 CHar), 7.41 (d, $^3J_{H,H}$ = 7.95 Hz, 2H, CHar), 7.33 (d, $^3J_{H,H}$ = 7.95 Hz, 1H, CHole), 6.63-6.59 (m, 1H, CHar), 2.40 (1H, CH_3) ^{13}C-NMR (125 MHz,

DMSO-d6):δ [ppm] = 164.3 (1C,CO), 140.1 (1C, CHole), 138.8 (1C, Cq), 134.7 (1C, Cq), 131.4 (1C, CHar), 130.4 (2C, 2 x CHar), 127.2 (1C, Cq), 126.5 (2C, 2 x CHar), 118.0 (1C, CHar), 117.1 (1C, Cq), 112.9 (1C, CHar), 94.5 (1C, CN), 20.5 (1C, CH_3).

(2*E*)-3-[1-(4-chlorophenyl)-1*H*-pyrrol-2-yl]-2-cyanoprop-2-enamide (68)

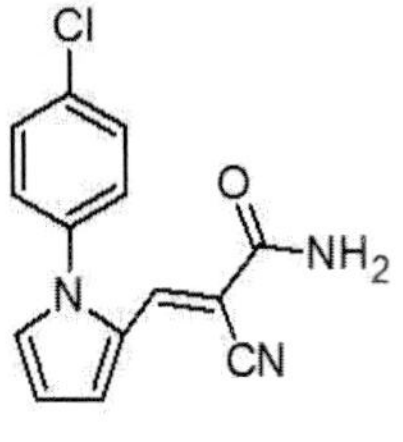

Cl N O NH_2 CN

Pale yellow crystals, 74% yield, Melting Point = 241.1 °C UV-Vis (MeOH: H_2O, 8:2,v/v):λ_{max} = 366 nm ^{1}H-NMR (500 MHz, DMSO-d6): δ [ppm] = 7.76 (s, 1H, CHole), 7.65-7.47 (m, 8H, 6 x CHar, 2 x NH), 6.61-6.59 (m, 1H, CHar) ^{13}C-NMR (125 MHz, DMSO-d6): δ [ppm] = 162.7 (1C, CO), 137.6 (1C, CHole), 136.2 (1C, Cq), 133.1 (1C, Cq), 130.4 (1C, CHar), 129.7 (2C, 2 x CHar), 128.5 (2C, 2 x CHar), 126.9 (1C, Cq), 117.4 (1C, Cq), 116.4 (1C, CHar), 112.2 (1C, CHar), 97.9 (1C, CN).

(2*E*)-3-[1-(4-chlorophenyl)-1*H*-pyrrol-2-yl]-2-cyanoprop-2-enoic acid (69)

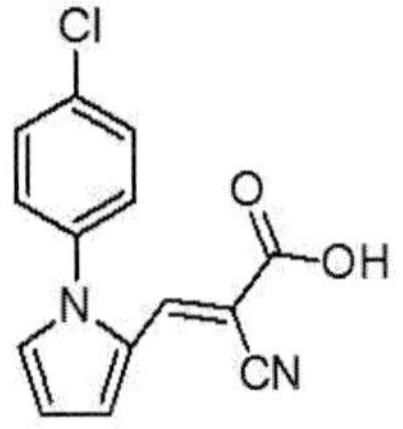

Cl N O OH CN

Yellow crystals, 74% yield, Melt-ing Point = 212.71 °C UV-Vis (MeOH: H_2O, 8:2,v/v):λ_{max} = 353 nm ^{1}H-NMR (500 MHz, DMSO-d6): δ [ppm] = 7.70-7.64 (m, 5H, CHole, 4 x CHar), 7.51 (d,$^3J_{H,H}$ = 8.40 Hz, 2H, 2 x CHar), 6.67-6.65 (m, 1H, CHar) ^{13}C-NMR (125 MHz, DMSO-d6): δ [ppm] = 164.1 (1C, CO), 139.8 (1C, CHole), 135.8 (1C, Cq), 133.7 (1C, Cq), 131.6 (1C, CHar), 129.7 (2C, 2 x CHar), 128.5 (2C, 2 x CHar), 126.9 (1C, Cq), 118.0 (1C, CHar), 117.1 (1C, Cq), 112.9 (1C, CHar), 94.5 (1C, CN).

(2*E*)-2-cyano-3-[1-(3,5-dichlorophenyl)-1*H*-pyrrol-2-yl]prop-2-enamide (70)

Pale yellow crystals, 77% yield, Melting Point = 287.6 °C UV-Vis (MeOH: H_2O, 8:2,v/v):λ_{max} = 363 nm

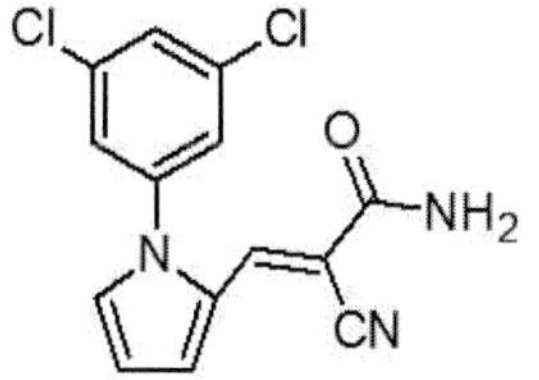

^{1}H-NMR (500 MHz, DMSO-d6): δ [ppm] = 7.83 (s, 1H, CHole), 7.70-7.54 (m, 7H, 5 x CHar, 2 x NH), 6.63-6.60 (m, 1H, CHar) ^{13}C-NMR (125 MHz, DMSO-d6): δ [ppm] = 162.7 (1C, CO), 139.3 (1C, Cq), 137.2 (1C, CHole), 134.6 (1C, Cq), 130.4 (1C, CHar), 128.3 (1C, CHar), 126.9 (1C, Cq), 125.7 (2C, 2 x CHar), 117.4 (1C, Cq), 117.1 (1C, CHar), 112.2 (1C, CHar), 98.5 (1C, CN).

(2*E*)-2-cyano-3-[1-(3,5-dichlorophenyl)-1*H*-pyrrol-2-yl]prop-2-enoic acid (71)

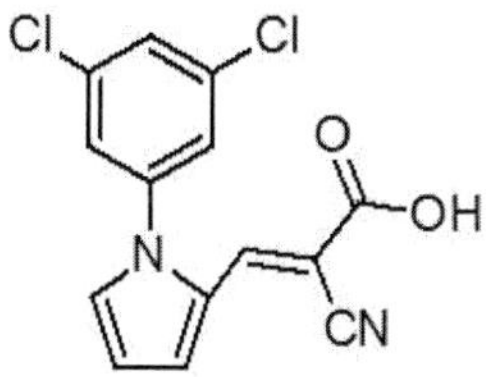

Yellow crystals, 79% yield, Melting Point = 275.3 °C UV-Vis (MeOH: H_2O, 8:2,v/v):λ_{max} = 351 nm ^{1}H-NMR (500 MHz, DMSO-d6): δ [ppm] = 7.85 (s, 1H, CHole), 7.72-7.66 (m, 5H, 5 x CHar), 6.69-6.65 (m, 1H, CHar) ^{13}C-NMR (125 MHz, DMSO-d6): δ [ppm] = 163.9 (1C, CO), 139.9 (1C, CHar), 139.2 (1C, Cq), 134.6 (1C, Cq), 131.7 (1C, CHar), 128.7 (1C, CHar), 126.9 (1C, Cq), 125.7 (2C, 2 x CHar), 118.4 (1C, CHar), 116.9 (1C, Cq), 112.9 (1C, CHar), 98.5 (1C, CN).

(2*E*)-2-cyano-3-[1-(4-hydroxyphenyl)-1*H*-pyrrol-2-yl]prop-2-enamide (72)

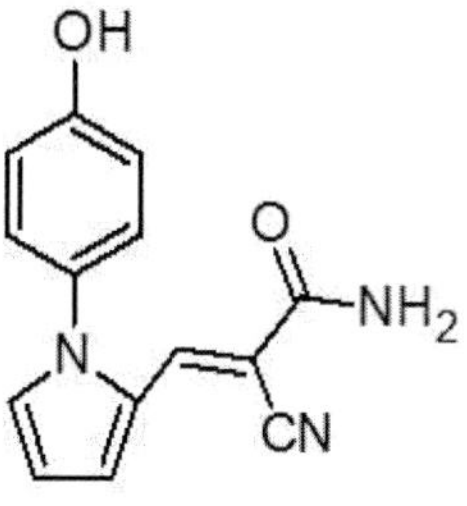

Yellow crystals, 86% yield, Melting Point = 230.1 °C UV-Vis (MeOH: H_2O, 8:2,v/v):λ_{max} = 371 nm ^{1}H-NMR (500 MHz, DMSO-d6): δ [ppm] = 7.67 (s, 1H, CHole), 7.54-7.43 (m, 4H, 2 x NH, 2 x CHar), 7.20 (d, $^3J_{H,H}$ = 8.45 Hz, 2H, 2 x CHar), 6.91 (d, $^3J_{H,H}$ = 8.45 Hz, 2H, 2 x CHar), 6.56-6.53 (m, 1H, CHar)

^{13}C-NMR (125 MHz, DMSO-d6): δ [ppm] = 162.7 (1C, CO), 157.8 (1C, Cq), 138.2 (1C, CHole), 130.7 (1C, CHar), 128.4 (1C, Cq), 128.1 (2C, 2 x CHar), 127.3 (1C, Cq), 117.5 (1C, Cq), 116.1 (1C, CHar), 115.9 (2C, 2 x CHar), 111.7 (1C, CHar), 96.6 (1C, CN).

(2*E*)-2-cyano-3-[1-(4-hydroxyphenyl)-1*H*-pyrrol-2-yl]prop-2-enoic acid (73)

Yellow crystals, 74% yield, Melting Point = 229.7 °C UV-Vis (MeOH: H_2O, 8:2,v/v):λ_{max} = 356 nm ^{1}H-NMR (500 MHz, DMSO-d6): δ [ppm] = 7.69 (s, 1H, CHole), 7.60-7.53 (m, 2H, 2 x CHar), 7.23 (d, $^3J_{H,H}$ = 8.45 Hz, 2H, 2 x CHar), 6.91 (d, $^3J_{H,H}$ = 8.45 Hz, 2H, 2 x CHar), 6.61-6.57 (m, 1H, CHar) ^{13}C-NMR (125 MHz, DMSO-d6): δ [ppm] = 164.1 (1C, CO), 157.8 (1C, Cq), 140.1 (1C, CHole), 131.7 (1C, CHar), 128.1 (2C, 2 x CHar), 127.1 (1C, Cq), 117.4 (1C, CHar), 117.1 (1C, Cq), 115.9 (2C, 2 x CHar), 112.4 (1C, CHar), 96.6 (1C, CN).

(2*E*)-2-cyano-3-[1-(4-methoxyphenyl)-1*H*-pyrrol-2-yl]prop-2-enamide (74)

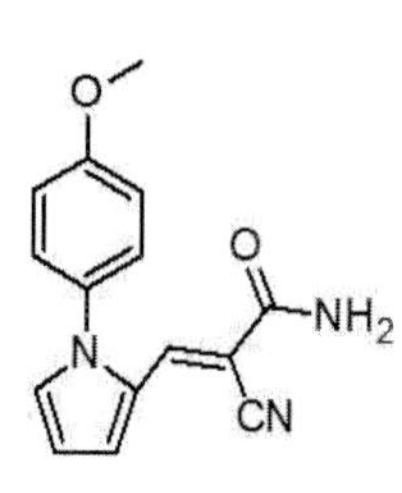

Yellow crystals, 89% yield, Melt-ing Point = 241.1 °C UV-Vis (MeOH: H_2O, 8:2,v/v):λ_{max} = 369 nm ^{1}H-NMR (500 MHz, DMSO-d6): δ [ppm] = 7.67 (s, 1H, CHole), 7.55-7.46 (m, 4H, 2 x NH, 2 x CHar), 7.34 (d, $^3J_{H,H}$ = 8.45 Hz, 2H, 2 x CHar), 7.11 (d, $^3J_{H,H}$ = 8.45 Hz, 2H, 2 x CHar), 6.57-6.55 (m, 1H, CHar), 3.84 (s, 3H, CH_3) ^{13}C-NMR (125 MHz, DMSO-d6): δ [ppm] = 162.9 (1C, CO), 159.1 (1C, Cq), 138.2 (1C, CHole), 130.7 (1C, CHar), 130.1 (1C, Cq), 127.9 (2C, 2 x CHar), 127.3 (1C, Cq), 116.3 (1C, CHar), 114.9 (2C, 2 x CHar), 111.7 (1C, CHar), 97.0 (1C, CN), 55.5 (1C, CH_3).

(2*E*)-2-cyano-3-[1-(4-methoxyphenyl)-1*H*-pyrrol-2-yl]prop-2-enoic acid (75)

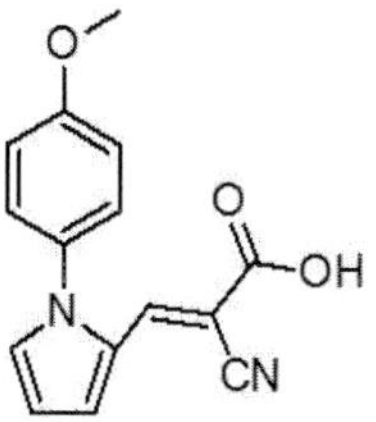

Yellow crystals, 66% yield, Melting Point = 245.8 °C UV-Vis (MeOH: H_2O, 8:2,v/v):λ_{max} = 355 nm ^{1}H-NMR (500 MHz, DMSO-d6): δ [ppm] = 7.69 (s, 1H, CHole), 7.63-7.56 (m, 2H, 2 x CHar), 7.38 (d, $^3J_{H,H}$ = 8.45 Hz, 2H, 2 x CHar), 7.11 (d, $^3J_{H,H}$ = 8.45 Hz, 2H, 2 x CHar), 6.62-6.60 (m, 1H, CHar), 3.85 (s, 3H, CH_3) ^{13}C-NMR (125 MHz, DMSO-d6): δ [ppm] = 162.9 (1C, CO), 159.1 (1C, Cq), 138.2 (1C, CHole), 130.7 (1C, CHar), 130.1 (1C, Cq), 128.2 (2C, 2 x CHar), 127.3 (1C, Cq), 116.3 (1C, CHar), 114.8 (2C, 2 x CHar), 111.7 (1C, CHar), 97.0 (1C, CN), 55.5 (1C, CH_3).

(2*E*)-2-cyano-3-(1-phenyl-1*H*-pyrrol-2-yl)prop-2-enamide (76)

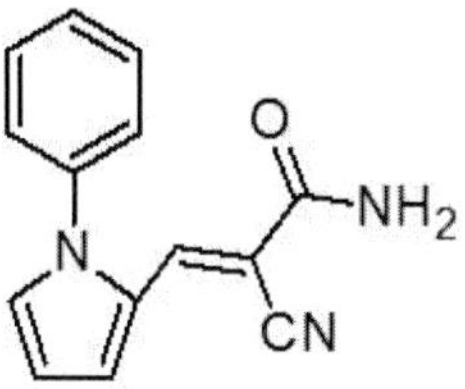

Pale yellow crystals, 69% yield, Melting Point = 204.7 °C UV-Vis (MeOH: H_2O, 8:2,v/v):λ_{max} = 367 nm ^{1}H-NMR (500 MHz, DMSO-d6): δ [ppm] = 7.71 (s, 1H, CHole), 7.61-7.55 (m, 7H, 2 x NH, 5 x CHar), 7.43 (d, $^3J_{H,H}$ = 7.30 Hz, 2H, 2 x CHar), 6.60 (t, $^3J_{H,H}$ = 2.80 Hz, 1H, CHar) ^{13}C-NMR (125 MHz, DMSO-d6): δ [ppm] = 162.9 (1C, CO), 138.0 (1C, CHole), 137.3 (1C, Cq), 130.4 (1C, CHar), 129.6 (2C, 2 x CHar), 128.6 (1C, CHar), 126.9 (1C, Cq), 126.7 (2C, 2 x CHar), 117.5 (1C, Cq), 116.5 (1C, CHar), 111.7 (1C, CHar), 112.0 (1C, CHar), 97.3 (1C, CN).

(2*E*)-2-cyano-3-(1-phenyl-1*H*-pyrrol-2-yl)prop-2-enoic acid (77)

Yellow crystals, 72% yield, Melt-ing Point = 195.3 °C UV-Vis (MeOH: H_2O, 8:2,v/v):λ_{max} = 3541 nm ^{1}H-NMR (500 MHz, DMSO-d6): δ [ppm] = 7.73 (s, 1H, CHole), 7.67-7.55 (m, 5H, 5 x CHar), 7.46 (d, $^3J_{H,H}$ = 7.45 Hz, 2H, 2 x CHar),

6.65 (t, $^3J_{H,H}$ = 2.80 Hz, 1H, CHar) ^{13}C-NMR (125 MHz, DMSO-d6): δ [ppm] = 164.0 (1C, CO), 140.1 (1C, CHole), 136.9 (1C, Cq), 131.7 (1C, CHar), 129.6 (2C, 2 x CHar), 128.9 (1C, CHar), 126.9 (1C, Cq), 126.7 (2C, 2 x CHar), 117.9 (1C, CHar), 117.1 (1C, Cq), 112.7 (1C, CHar), 97.3 (1C, CN).

(2*E*)-3-[1-(2H-1,3-benzodioxol-5-yl)-1*H*-pyrrol-2-yl]-2-cyanoprop-2-enamide (78)

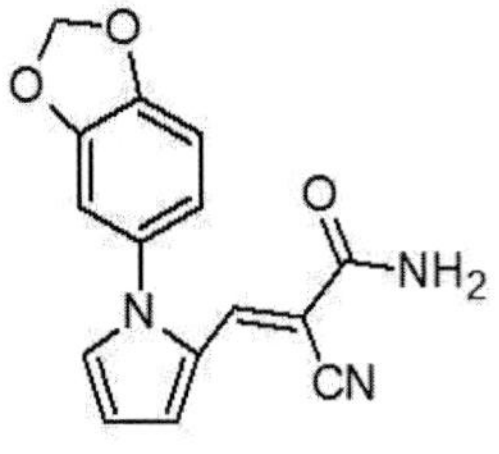

O O N O NH_2 CN

Yellow crystals, 71% yield, Melting Point = 208.3 °C UV-Vis (MeOH: H_2O, 8:2,v/v):λ_{max} = 366 nm ^{1}H-NMR (500 MHz, DMSO-d6): δ [ppm] = 7.35 (s, 1H, CHole), 7.16-6.88 (m, 7H, 2 x NH, 5 x CHar), 6.41 (t, $^3J_{H,H}$ = 2.80 Hz, 1H, CHar), 6.12 (s, 2H, CH_2) ^{13}C-NMR (125 MHz, DMSO-d6): δ [ppm] = 163.9 (1C, CO), 147.8 (1C, CHole), 147.3 (1C, Cq), 131.4 (1C, CHar), 130.6 (1C, Cq), 127.3 (1C, Cq), 120.6 (2C, 2 x CHar), 117.5 (1C, Cq), 112.5 (1C, CHar), 108.3 (1C, CHar), 108.1 (1C, CHar), 103.1 (1C, CH_2), 102.2 (1C, CN).

(2*E*)-3-[1-(2H-1,3-benzodioxol-5-yl)-1*H*-pyrrol-2-yl]-2-cyanoprop-2-enoic acid (79)

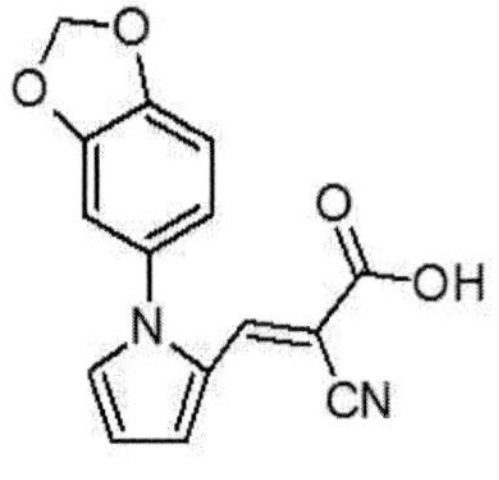

O O N O OH CN

Yellow crystals, 82% yield, Melting Point = 217.3 °C UV-Vis (MeOH: H2O, 8:2,v/v):λ_{max} = 354 nm ^{1}H-NMR (500 MHz, DMSO-d6): δ [ppm] = 7.73 (s, 1H, CHole), 7.61-6.55 (m, 2H, 2 x CHar), 7.14-7-08 (m, 2H, 2 x CHar), 6.91 (d, $^3J_{H,H}$ = 8.40 Hz, 1H, CHar), 6.60 (m, 1H, CHar), 6.17 (s, 2H, CH_2) ^{13}C-NMR (125 MHz, DMSO-d6): δ [ppm] = 163.9 (1C, CO), 147.8 (1C, CHole), 147.3 (1C, Cq), 131.4 (1C, CHar), 130.6 (1C, Cq), 127.3 (1C, Cq), 120.6 (2C, 2 x CHar), 117.5 (1C, Cq), 112.5 (1C, CHar), 108.3 (1C, CHar), 108.1 (1C, CHar), 103.1 (1C, CH2), 102.2 (1C, CN).

(2*E*)-2-cyano-3-[1-(1,3-thiazol-2-yl)-1*H*-pyrrol-2-yl]prop-2-enamide (80)

Brown crystals, 87% yield, Melting Point = 186.1 °C UV-Vis (MeOH: H_2O, 8:2,v/v):λ_{max} = 357 nm ^{1}H-NMR (500 MHz, DMSO-d6): δ [ppm] = 8.52 (s, 1H, CHole), 7.85-7.77 (m, 3H, 3 x CHar), 7.71 (s, br, 1H, NH), 7.65 (s, br, 1H, NH), 7.58 (s, 1H, CHar), 6.68-66 (m, 1H, CHar) ^{13}C-NMR (125 MHz, DMSO-d6): δ [ppm] = 162.0 (1C, CO), 158.4 (1C, Cq), 140.7 (1C, CHole), 138.4 (1C, CHar), 129.5 (1C, CHar), 127.3 (1C, Cq), 119.9 (1C, CHar), 118.3 (1C, CHar), 117.2 (1C, Cq), 113.4 (1C, CHar), 100.1 (1C, CN).

(2*E*)-2-cyano-3-[1-(1,3-thiazol-2-yl)-1*H*-pyrrol-2-yl]prop-2-enoic acid (81)

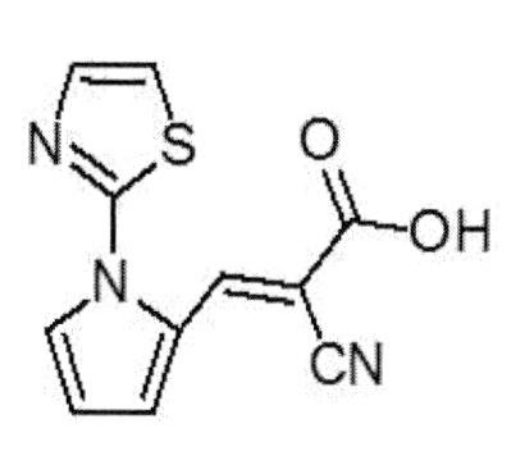

Yellow crystals, 78% yield, Melt-ing Point = 233.8 °C UV-Vis (MeOH: H_2O, 8:2,v/v):λ_{max} = 345 nm ^{1}H-NMR (500 MHz, DMSO-d6): δ [ppm] = 8.70 (s, 1H, CHole), 7.85-7.69 (m, 4H, 4 x CHar), 7.71-7.68 (m, 1H, CHar) ^{13}C-NMR (125 MHz, DMSO-d6): δ [ppm] = 163.7 (1C, CO), 158.4 (1C, Cq), 140.7 (1C, CHole), 138.4 (1C, CHar), 130.6 (1C, CHar), 127.1 (1C, Cq), 120.2 (1C, CHar), 118.3 (1C, CHar), 117.2 (1C, Cq), 114.0 (1C, CHar), 100.1 (1C, CN).

(2*E*)-2-cyano-3-(1-methyl-1*H*-pyrrol-2-yl)prop-2-enamide (82)

Greenish yellow crystals, 69% yield, Melting Point = 180.1 °C UV-Vis (MeOH: H_2O, 8:2,v/v):λ_{max} = 374 nm ^{1}H-NMR (500 MHz, DMSO-d6): δ [ppm] = 7.98 (s, 1H, CHole), 7.71 (s, br, 1H, NH), 7.49 (s, br, 1H, NH), 7.38 (d, $^3J_{H,H}$ = 3.80 Hz, 1H, CHar), 7.30 (s, 1H, CHar), 6.35-6.33 (m, 1H,

CHar), 3.78 (s, 3H, CH_3) ^{13}C-NMR (125 MHz, DMSO-d6): δ [ppm] = 163.7 (1C, CO), 136.6 (1C, CHole), 130.7 (1C, CHar), 126.5 (1C, Cq), 117.9 (1C, Cq), 116.1 (1C, CHar), 110.7 (1C, CHar), 96.1 (1C, CN), 33.7 (1C, CH_3).

(2*E*)-2-cyano-3-(1-methyl-1*H*-pyrrol-2-yl)prop-2-enoic acid (83)

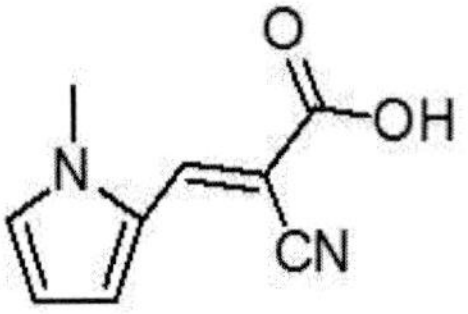

Yellow crystals, 81% yield, Melting Point = 235.6 °C UV-Vis (MeOH: H_2O, 8:2,v/v):λ_{max} = 360 nm ^{1}H-NMR (500 MHz, DMSO-d6): δ [ppm] = 8.06 (s, 1H, CHole), 7.47 (d, $^3J_{H,H}$ = 4.00 Hz, 1H, CHar), 7.40 (s, 1H, CHar), 6.41-6.39 (m, 1H, CHar), 3.80 (s, 3H, CH_3) ^{13}C-NMR (125 MHz, DMSO-d6): δ [ppm] = 164.5 (1C, CO), 139.6 (1C, CHole), 132.7 (1C, CHar), 126.5 (1C, Cq), 117.7 (1C, CHar), 117.5 (1C, Cq), 111.3 (1C, CHar), 92.4 (1C, CN), 33.7 (1C, CH_3).

(2*E*)-2-cyano-3-[1-(4-fluorophenyl)-1*H*-pyrrol-2-yl]prop-2-enamide (84)

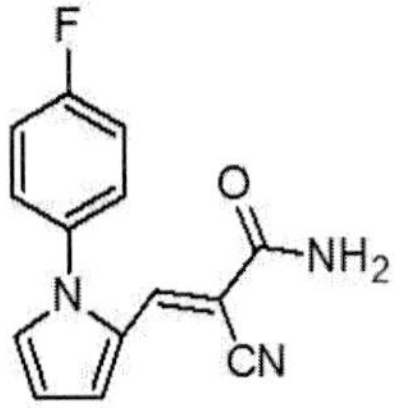

Yellow crystals, 81% yield, Melting Point = 190.6 °C UV-Vis (MeOH: H_2O, 8:2,v/v):λ_{max} = 366 nm ^{1}H-NMR (500 MHz, DMSO-d6): δ [ppm] = 7.65 (s, 1H, CHole), 7.61-7.42 (m, 8H, 6 x CHar, 2 x NH), 6.60-6.59 (m, 1H, CHar) ^{13}C-NMR (125 MHz, DMSO-d6): δ [ppm] = 163.0 (1C, CO), 137.6 (1C, CHole), 133.5 (1C, Cq), 130.5 (1C, CHar), 129.0 (1C, CHar), 127.5 (1C, Cq), 117.6 (1C, Cq), 116.3 (2C, 2 x CHar), 112.2 (2C, 2 x CHar), 97.5 (1C, CN).

(2*E*)-2-cyano-3-[1-(4-fluorophenyl)-1*H*-pyrrol-2-yl]prop-2-enoic acid (85)

Yellow crystals, 79% yield, Melting Point = 181.8 °C UV-Vis (MeOH: H_2O, 8:2,v/v):λ_{max} = 349 nm ^{1}H-NMR (500 MHz, DMSO-d6): δ [ppm] = 7.66 (s, 1H, CHole), 7.63-7.44 (m, 6H, 6 x CHar), 6.64-6.62 (m, 1H, CHar) ^{13}C-NMR (125 MHz, DMSO-d6): δ [ppm] = 163.0 (1C, CO), 137.6 (1C, CHole), 133.5 (1C, Cq), 130.5 (1C, CHar), 129.0 (1C, CHar), 127.5 (1C, Cq), 117.6 (1C, Cq), 116.3 (2C, 2 x CHar), 112.2 (2C, 2 x CHar), 97.5 (1C, CN).

(2*E*)-2-cyano-3-(1-methyl-1*H*-pyrrol-3-yl)prop-2-enoic acid (87)

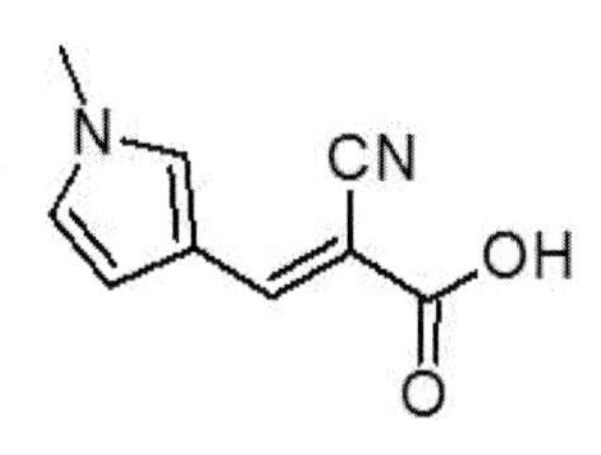

Brown crystals, 76% yield, Melt-ing Point = 178.9 °C UV-Vis (MeOH: H_2O, 8:2,v/v):λ_{max} = 331 nm ^{1}H-NMR (500 MHz, DMSO-d6): δ [ppm] = 13.15 (s, br, 1H, COOH), 11.84 (s, 1H, NH), 8.21 (s, 1H, CHole), 7.72 (s, 1H, CHar), 7.04 (s, 1H, CHar), 6.91 (s, 1H, CHar) ^{13}C-NMR (125 MHz, DMSO-d6): δ [ppm] = 164.6 (1C, CO), 139.7 (1C, CHole), 129.6 (1C, CHar), 122.0 (1C, CHar), 118.1 (1C, Cq), 117.5 (1C, Cq), 107.3 (1C, CHar), 94.0 (1C, CN).

(2*E*)-2-cyano-3-(5-cyclopropylcyclopenta-1,3-dien-1-yl)prop-2-enamide (88)

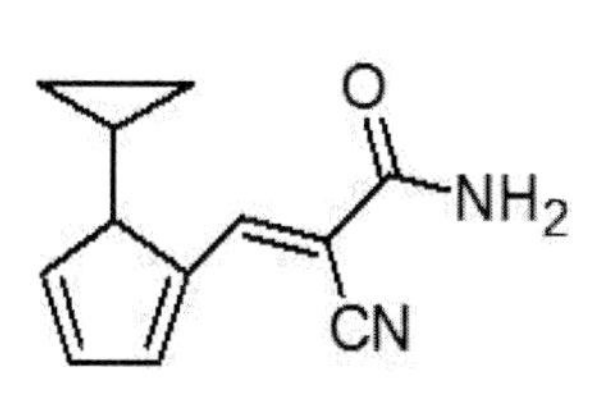

Yellow crystals, 84% yield, Melting Point = 215.1 °C UV-Vis (MeOH: H_2O, 8:2,v/v):λ_{max} = 373 nm ^{1}H-NMR (500 MHz, DMSO-d6): δ [ppm] = 13.41 (s, br, 1H, COOH), 8.42 (s, 1H, CHole), 7.42 (d, $^3J_{H,H}$ = 8.45 Hz, 2H, 2 x CHar), 6.35 (s, 1H, CHar), 3.55 (s, 1H, CH), 1.07 (s, 2H, CH_2), 0.99 (s, 2H, CH_2) ^{13}C-NMR (125 MHz, DMSO-d6): δ [ppm] = 164.2 (1C, CO), 139.7 (1C, CHole), 131.2 (1C, CHar), 127.6 (1C, Cq), 118.1 (1C, CHar), 117.3 (1C, Cq), 111.3 (1C, CHar), 93.0 (1C, CN), 27.7 (2C, 2 x CH_2).

(2*E*)-3-(1-tert-butyl-1*H*-pyrrol-3-yl)-2-cyanoprop-2-enamide (89)

White crystals, 67% yield, Melting Point = 176.8 °C UV-Vis (MeOH: H_2O, 8:2,v/v):λ_{max} = 362 nm ^{1}H-NMR (500 MHz, DMSO-d6): δ [ppm] = 8.04 (s, 1H, CHole), 7.72 (s, 1H, CHar), 7.49, (s, br, 1H, NH), 7.41 (s, br, 1H, NH), 7.22 (s, 1H, CHar), 6.85 (s, 1H, CHar), 1.50 (9H, 3 x CH_3) ^{13}C-NMR (125 MHz, DMSO-d6): δ [ppm] = 164.2 (1C, CO), 146.6 (1C, CHole), 128.2 (1C, CHar), 122.3 (1C, CHar), 118.7 (1C, Cq), 117.5 (1C, Cq), 108.3 (1C, CHar), 97.5 (1C, CN), 56.5 (1C, Cq), 30.5 (3C, 3 x CH_3).

(2*E*)-3-(1-tert-butyl-1*H*-pyrrol-3-yl)-2-cyanoprop-2-enoic acid (90)

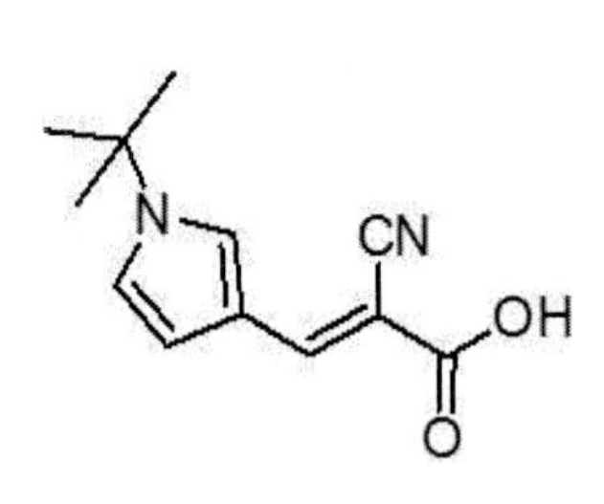

White crystals, 76% yield, Melting Point = 186.3 °C UV-Vis (MeOH: H_2O, 8:2,v/v):λ_{max} = 355 nm ^{1}H-NMR (500 MHz, DMSO-d6): δ [ppm] = 13.16 (s, 1H, COOH), 8.13 (s, 1H, CHole), 7.89 (s, 1H, CHar), 7.27 (s, 1H, CHar), 6.91 (s, 1H, CHar), 1.54 (9H, 3 x CH_3) ^{13}C-NMR (125 MHz, DMSO-d6): δ [ppm] = 165.1 (1C, CO), 1469.7 (1C, CHole), 130.2 (1C, CHar), 123.3 (1C, CHar), 118.2 (1C, Cq), 108.3 (1C, CHar), 97.5 (1C, CN), 56.5 (1C, Cq), 30.5 (3C, 3 x CH_3).

7

Appendix 1

X-ray Crystallography

Single compound crystals were grown at room temperature using a saturated acetone and water (7:3, v/v) stock solution. The crystallographic measurements were carried out (by Mr. Rominger, University of Heidelberg, Germany) on a Bruker APEX-II Quazar (matrices 19 and 45), Bruker APEX-II CCD (matrix 10) or a BRUKER APEX diffractometer (matrices 11, 22, 26). Intensities were corrected for Lorentz and polarisation effects, an empirical absorption corrections were applied using SADABS [1] based on the Laue symmetry of the reciprocal space. Structures were refined against F2 with a Full-matrix least-squares algorithm using the SHELXL-2014/7 software [2]. Crystal structure visualization and crystal packing calculation was done using Mercury 3.7 (build RC1), (Cambridge Crystallographic Data Centre, UK). CCDC 1484580 (matrix 10), 1484581 (matrix 11), 1484582 (matrix 19), 1484583 (matrix 22), 1484584 (matrix 26), 1484585 (matrix 45) contain the supplementary crystallographic data for this paper. These data can be obtained free of charge from The Cambridge Crystallographic Data Centre via
www.ccdc.cam.ac.uk/dataR equest/cif.

Crystallography:

Details of the single crystal X-ray crystallography for the following matrices

1) **Matrix 10**

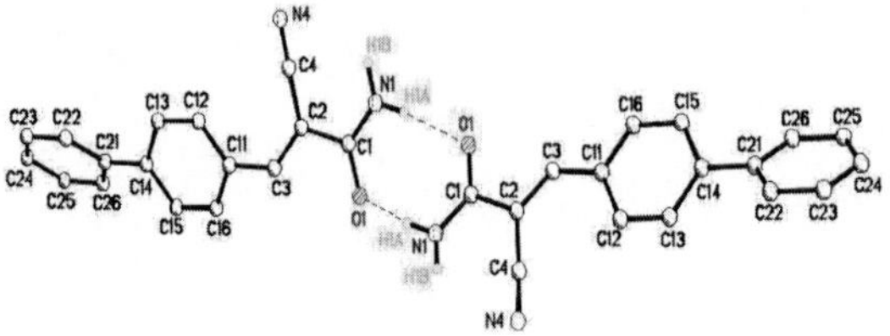

Crystal data and structure refinement

Identification code	hdj6	
Empirical formula	$C_{16.75}H_{13.50}N_{2}O_{1.25}$	
Formula weight	262.79	
Temperature	200(2) K	
Wavelength	0.71073 Å	
Crystal system	triclinic	
Space group	$P\bar{1}$	
Z	4	
Unit cell dimensions	a = 9.123(2) Å	α= 84.270(6) deg.
	b = 10.860(3) Å	β = 89.011(6) deg.
	c = 13.842(3) Å	γ = 82.514(6) deg.
Volume	1352.9(5) $Å^3$	
Density (calculated)	1.29 g/cm^3	
Absorption coefficient	0.08 mm^{-1}	
Crystal shape	polyhedron	
Crystal size	0.170 x 0.160 x 0.150 mm^3	
Crystal colour	colourless	
Theta range for data collection	1.9 to 21.7 deg.	
Index ranges	$-9 \le h \le 9$, $-11 \le k \le 11$, $-14 \le l \le 14$	
Reflections collected	13018	
Independent reflections	3167 (R(int) = 0.0665)	
Observed reflections	2181 (I > 2σ(I))	
Absorption correction	Semi-empirical from equivalents	
Max. and min. transmission	0.96 and 0.79	
Refinement method	Full-matrix least-squares on F^2	
Data/restraints/parameters	3167 / 412 / 395	
Goodness-of-fit on F^2	1.11	
Final R indices (I>2sigma(I))	R1 = 0.071, wR2 = 0.158	
Largest diff. peak and hole	0.22 and -0.27 $eÅ^{-3}$	

2) **Matrix 22**

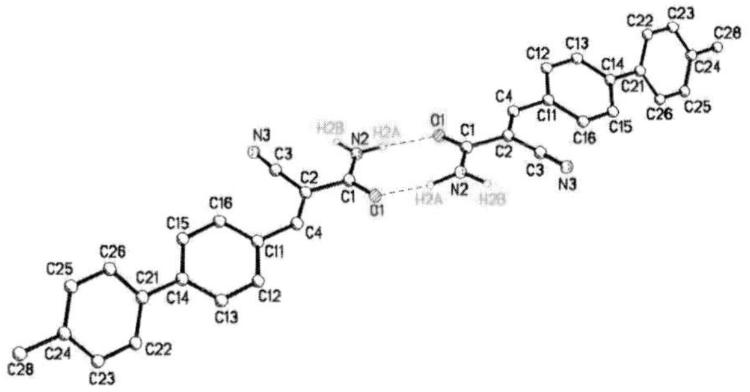

Crystal data and structure refinement.

Identification code	hdj2		
Empirical formula	$C_{17}H_{14}N_2O$		
Formula weight	262.30		
Temperature	200(2) K		
Wavelength	0.71073 Å		
Crystal system	triclinic		
Space group	P 1		
Z	4		
Unit cell dimensions	a = 9.8273(13) Å	α = 90.213(4)	deg.
	b = 10.8890(17) Å	β = 110.004(4)	deg.
	c = 13.5247(19) Å	γ = 90.524(5)	deg.
Volume	1359.9(3) $Å^3$		
Density (calculated)	1.28 g/cm^3		
Absorption coefficient	0.08 mm^{-1}		
Crystal shape	parallelepiped		
Crystal size	0.180 x 0.170 x 0.130 mm^3		
Crystal colour	yellow		
Theta range for data collection	1.6 to 24.2 deg.		
Index ranges	$-6 \leq h \leq 9$, $-9 \leq k \leq 12$, $-15 \leq l \leq 15$		
Reflections collected	8248		
Independent reflections	4296 (R(int) = 0.0334)		
Observed reflections	2835 ($I > 2\sigma(I)$)		
Absorption correction	Semi-empirical from equivalents		
Max. and min. transmission	0.96 and 0.90		
Refinement method	Full-matrix least-squares on F^2		
Data/restraints/parameters	4296 / 0 / 379		
Goodness-of-fit on F^2	1.03		
Final R indices (I>2sigma(I))	R1 = 0.067, wR2 = 0.138		
Largest diff. peak and hole	0.38 and -0.26 $eÅ^{-3}$		

3) Matrix 11

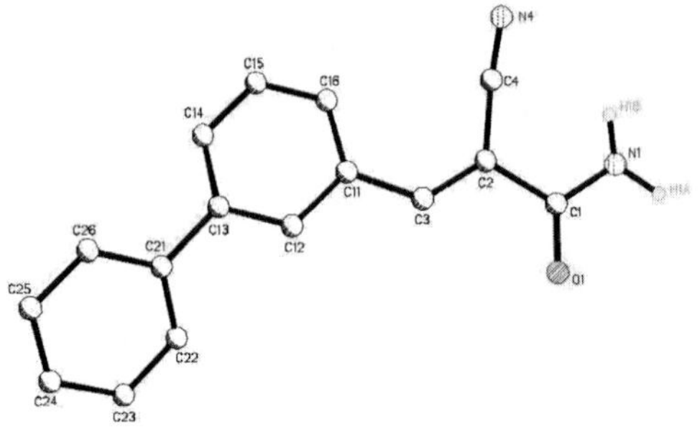

Crystal data and structure refinement.

Identification code	hdj9	
Empirical formula	$C_{16}H_{12}N_2O$	
Formula weight	248.28	
Temperature	200(2) K	
Wavelength	0.71073 Å	
Crystal system	triclinic	
Space group	P 1	
Z	4	
Unit cell dimensions	a = 9.924(3) Å	α = 86.187(9) deg.
	b = 10.552(4) Å	β = 89.457(7) deg.
	c = 12.185(3) Å	γ = 83.500(9) deg.
Volume	1265.0(6) $Å^3$	
Density (calculated)	1.30 g/cm^3	
Absorption coefficient	0.08 mm^{-1}	
Crystal shape	Keil	
Crystal size	0.210 x 0.090 x 0.060 mm^3	
Crystal colour	colourless	
Theta range for data collection	1.9 to 20.2 deg.	
Index ranges	$-9 \le h \le 9$, $-10 \le k \le 7$, $-11 \le l \le 9$	
Reflections collected	3235	
Independent reflections	2382 (R(int) = 0.0362)	
Observed reflections	1273 ($I > 2\sigma(I)$)	
Absorption correction	Semi-empirical from equivalents	
Max. and min. transmission	0.96 and 0.76	
Refinement method	Full-matrix least-squares on F^2	
Data/restraints/parameters	2382 / 322 / 359	
Goodness-of-fit on F^2	1.06	
Final R indices (I>2sigma(I))	R1 = 0.072, wR2 = 0.145	
Largest diff. peak and hole	0.22 and -0.25 $eÅ^{-3}$	

4) Matrix 26

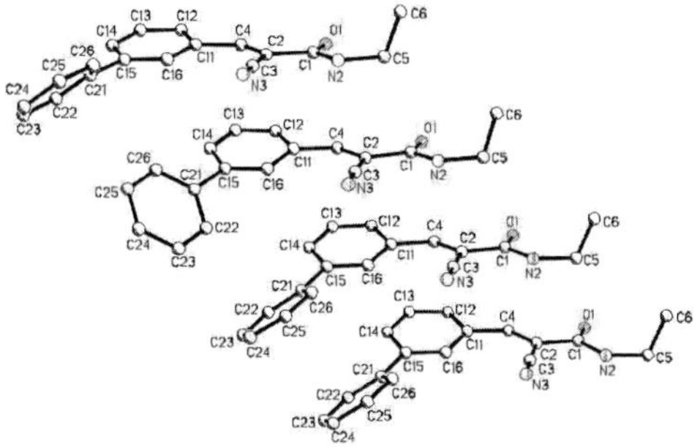

Crystal data and structure refinement for (26).

Identification code	hdj3	
Empirical formula	$C_{18}H_{16}N_2O$	
Formula weight	276.33	
Temperature	200(2) K	
Wavelength	0.71073 Å	
Crystal system	triclinic	
Space group	$P\bar{1}$	
Z	8	
Unit cell dimensions	a = 12.034(7) Å	α = 63.781(11) deg.
	b = 16.341(10) Å	β = 77.633(11) deg.
	c = 17.484(11) Å	γ = 80.249(11) deg.
Volume	3002(3) $Å^3$	
Density (calculated)	1.22 g/cm^3	
Absorption coefficient	0.08 mm^{-1}	
Crystal shape	needle	
Crystal size	0.200 x 0.190 x 0.140 mm^3	
Crystal colour	colourless	
Theta range for data collection	1.3 to 22.8 deg.	
Index ranges	-12≤h≤12, -17≤k≤17, -18≤l≤18	
Reflections collected	17222	
Independent reflections	7785 (R(int) = 0.0675)	
Observed reflections	3267 (I > 2σ(I))	
Absorption correction	Semi-empirical from equivalents	
Max. and min. transmission	0.96 and 0.79	
Refinement method	Full-matrix least-squares on F^2	
Data/restraints/parameters	7785 / 1985 / 819	
Goodness-of-fit on F^2	1.05	
Final R indices (I>2sigma(I))	R1 = 0.134, wR2 = 0.323	
Largest diff. peak and hole	0.63 and -0.40 $eÅ^{-3}$	

5) Matrix 19

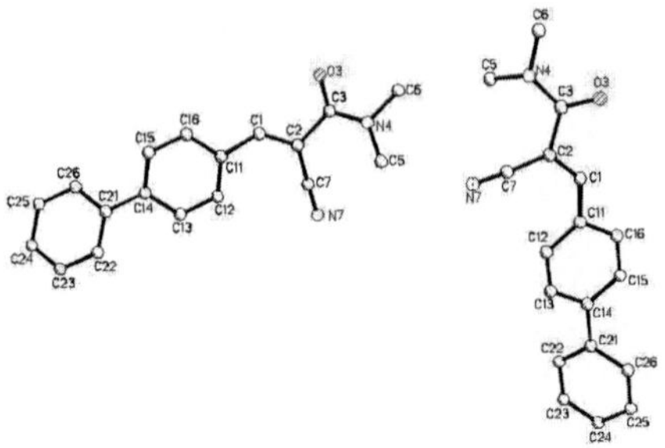

Crystal data and structure refinement for (19).

Identification code	hdj13	
Empirical formula	$C_{18}H_{16}N_2O_2$	
Formula weight	276.33	
Temperature	200(2) K	
Wavelength	0.71073 Å	
Crystal system	triclinic	
Space group	$P\bar{1}$	
Z	4	
Unit cell dimensions	a = 7.8375(9) Å	α = 86.735(3) deg.
	b = 9.2459(10) Å	β = 82.286(3) deg.
	c = 22.239(3) Å	γ = 64.911(3) deg.
Volume	1446.3(3) $Å^3$	
Density (calculated)	1.27 g/cm^3	
Absorption coefficient	0.08 mm^{-1}	
Crystal shape	needle	
Crystal size	1.11 x 0.11 x 0.05 mm^3	
Crystal colour	colourless	
Theta range for data collection	0.9 to 25.1 deg.	
Index ranges	-9≤h≤9, -9≤k≤10, -26≤l≤25	
Reflections collected	9213	
Independent reflections	4972 (R(int) = 0.0293)	
Observed reflections	3567 (I > 2σ(I))	
Absorption correction	Semi-empirical from equivalents	
Max. and min. transmission	0.96 and 0.85	
Refinement method	Full-matrix least-squares on F^2	
Data/restraints/parameters	4972 / 0 / 384	
Goodness-of-fit on F^2	1.01	
Final R indices (I>2sigma(I))	R1 = 0.064, wR2 = 0.163	
Largest diff. peak and hole	0.55 and -0.23 $eÅ^{-3}$	

6) **Matrix 45**

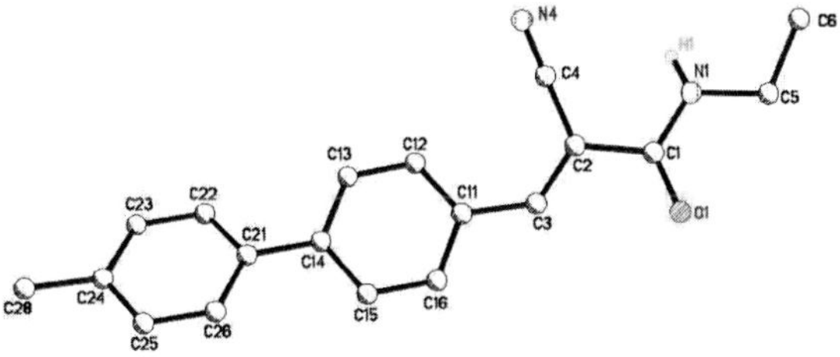

Crystal data and structure refinement for (45).

Identification code	hdj5	
Empirical formula	$C_{19}H_{18}N_{2}O$	
Formula weight	290.35	
Temperature	200(2) K	
Wavelength	0.71073 Å	
Crystal system	orthorhombic	
Space group	$Pca2_1$	
Z	8	
Unit cell dimensions	a = 28.6884(13) Å	α =90 deg.
	b = 11.0433(5) Å	β =90 deg.
	c = 9.6341(4) Å	γ=90 deg.
Volume	3052.2(2) $Å^3$	
Density (calculated)	1.26 g/cm^3	
Absorption coefficient	0.08 mm^{-1}	
Crystal shape	needle	
Crystal size	0.180 x 0.090 x 0.090 mm^3	
Crystal colour	colourless	
Theta range for data collection	1.4 to 24.9 deg.	
Index ranges	-33≤h≤33, -13≤k≤12, -10≤l≤11	
Reflections collected	23526	
Independent reflections	5070 (R(int) = 0.0384)	
Observed reflections	4346 (I > 2σ(I))	
Absorption correction	Semi-empirical from equivalents	
Max. and min. transmission	0.96 and 0.83	
Refinement method	Full-matrix least-squares on F^2	
Data/restraints/parameters	5070 / 1 / 409	
Goodness-of-fit on F^2	1.1 1	
Final R indices (I>2sigma(I))	R1 = 0.041, wR2 = 0.100	
Absolute structure parameter	-0.5(7)	
Largest diff. peak and hole	0.15 and -0.17 $eÅ^{-3}$	

References

1. G. M. Sheldrick, program SADABS 2012/1 for absorption correction. Bruker Analytical X-ray-Division, Madison, Wisconsin 2012
2. Sheldrick GM. Crystal structure refinement with SHELXL. ActaCryst. 2015;C17,3-8

8

Appendix 2

Physicochemical properties of the phenylcinnamic acid derivatives

Nr	Structure	logP	pKa	HD	HA	Monoisotopic mass	λmax	ελmax	ελ355nm	ελ337 nm	MP °C
1	O, NH_2, CN, O, O	2.41	10.01	1	4	292.2940	323	26047	8648	21447	253.4
2	O, O, NH, NH_2, CN	2.44	7.83	2	3	291.3092	342	24184	21615	23655	245.4
3	O, O, NH, NH_2, CN	2.44	7.83	2	3	291.3092	307	14144	1440	5033	208.7
4	HO, O, NH_2, CN	2.48	9.21	2	3	264.2836	299	20951	790	3311	204
5	OH, O, NH_2, CN	2.48	9.23	2	3	264.2836	338	6366	5718	6378	239.5
6	MeO, O, NH_2, CN	2.63	10.15	1	3	278.3104	341	26061	22926	25827	215.2
7	MeO, O, NH_2, CN	2.63	10.13	1	3	278.3104	335	26291	18956	26172	165.6
8	OMe, O, NH_2, CN	2.63	9.96	1	3	278.3104	339	20662	17529	20552	194.1

Nr	Structure	logP	pKa	HD	HA	Monoisotopic mass	λmax	ελmax	ελ355nm	ελ337 nm	MP °C
9	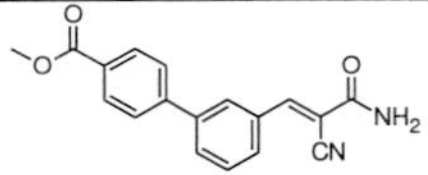	2.79	10.1	1	3	306.3208	276	30576	236	2960	208.9
10	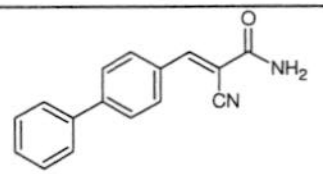	2.79	10.26	1	2	248.2842	333	28320	18144	27958	173.1
11	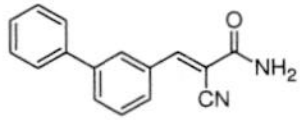	2.79	10.38	1	2	248.2842	300	19283	967	3751	153.4
12	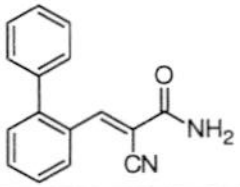	2.79	10.46	1	2	248.2842	301	13720	257	3196	139.9
13	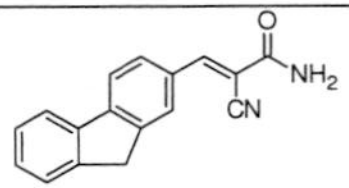	2.91	10.26	1	2	260.2952	350	29620	28780	26043	195.6
14	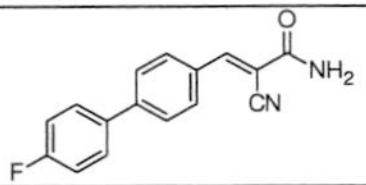	2.93	10.07	1	2	266.2746	334	26793	16994	26461	200.9
15	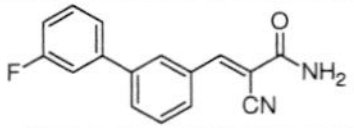	2.93	9.52	1	2	266.2746	298	15239	396	2684	120.6
16	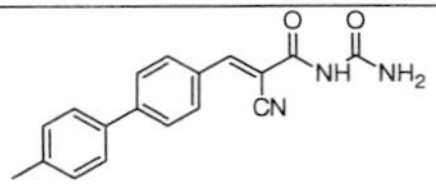	2.96	7.83	2	3	305.3361	300	33564	1369	5773	248.3

Nr	Structure	logP	pKa	HD	HA	Monoisotopic mass	λmax	ελmax	ελ355nm	ελ337 nm	MP °C
17	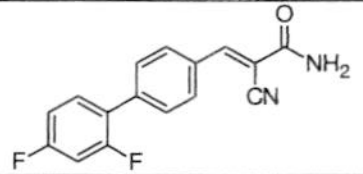 O NH2 CN F F	3.07	9.27	1	2	284.2651	325	23971	8383	20450	236.3
18	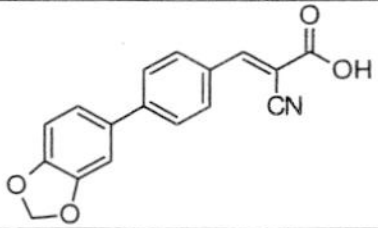 O OH CN O O	3.22	2.55	1	5	293.2787	343	21998	20244	21262	274
19	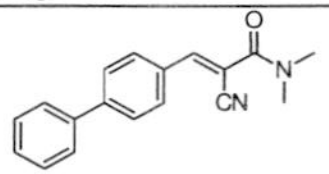 O N CN	3.24	0	0	2	276.3379	328	27044	11206	24398	125.3
20	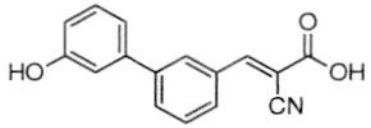 HO O OH CN	3.29	2.61	2	4	265.2683	293	16908	0	625	187.9
21	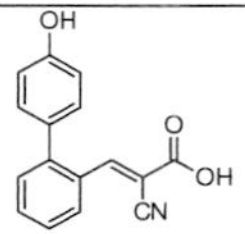 OH O OH CN	3.29	2.65	2	4	265.2683	288	12212	2995	5574	255.3
22	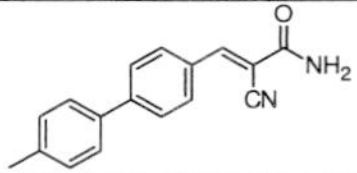 O NH2 CN	3.3	10.28	1	2	262.3110	340	29009	23932	28937	198.2
23	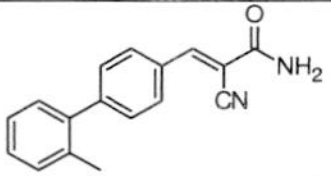 O NH2 CN	3.3	10.17	1	2	262.3110	323	23451	7795	19191	145.7
24	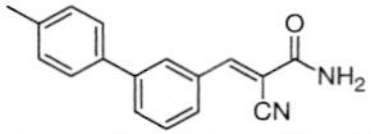 O NH2 CN	3.3	10.39	1	2	262.3110	302	17374	450	2560	154.1

Nr	Structure	logP	pKa	HD	HA	Monoisotopic mass	λmax	ελmax	ελ355nm	ελ337 nm	MP °C
25	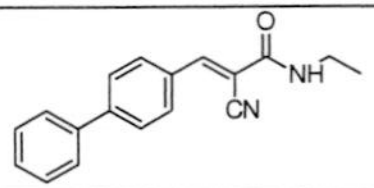 O NH CN	3.37	9.86	1	2	276.3379	327	27479	13450	24972	154.7
26	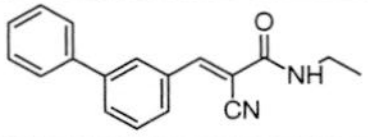 O NH CN	3.37	9.95	1	2	276.3379	300	21415	1892	5009	124.7
27	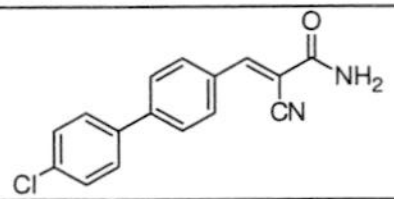 O NH_2 CN Cl	3.39	10.13	1	2	282.7292	335	31683	21306	31515	286.2
28	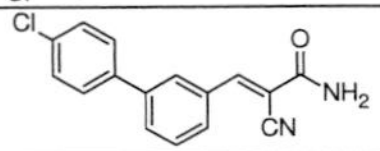 Cl O NH_2 CN	3.39	9.94	1	2	282.7292	300	18100	738	3113	157.4
29	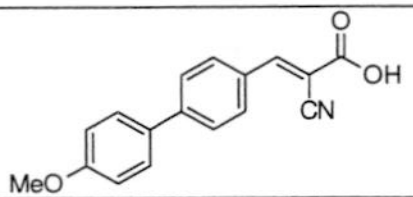 O OH CN MeO	3.44	2.69	1	4	279.2952	332	29654	19553	29118	252.6
30	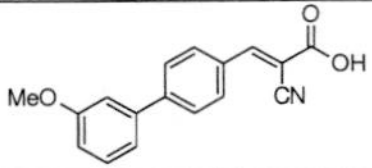 O OH CN MeO	3.44	2.67	1	4	279.2952	323	25209	8371	20760	198.1
31	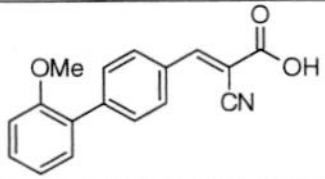 OMe O OH CN	3.44	2.5	1	4	279.2952	328	19948	26945	31592	197.8
32	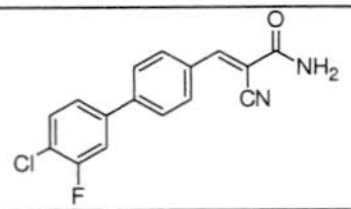 O NH_2 CN Cl F	3.53	9.69	1	2	300.7197	326	28166	12547	25126	217.7

Nr	Structure	logP	pKa	HD	HA	Monoisotopic mass	λmax	ελmax	ελ355nm	ελ337 nm	MP °C
33	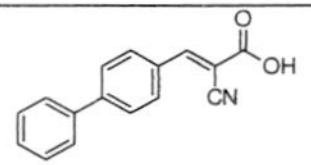 O OH CN	3.6	2.8	1	3	249.2689	325	28467	8549	23860	238.5
34	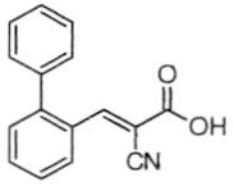 O OH CN	3.6	3	1	3	249.2689	295	13162	405	2867	184.2
35	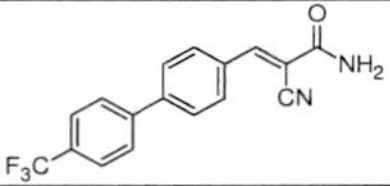 O NH_2 CN F_3C	3.67	10.14	1	2	316.282	326	31897	11614	28364	203.7
36	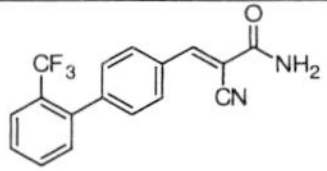 CF_3 O NH_2 CN	3.67	9.27	1	2	316.2824	312	26301	929	9817	138.6
37	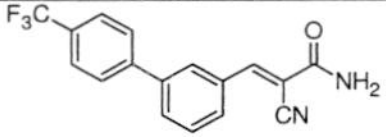 F_3C O NH_2 CN	3.67	10.19	1	2	316.2824	297	17569	585	2903	142.9
38	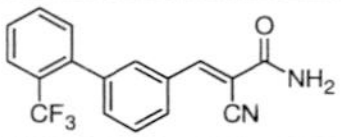 O NH_2 CF_3 CN	3.67	8.68	1	2	316.2824	298	45775	0	3330	117.2
39	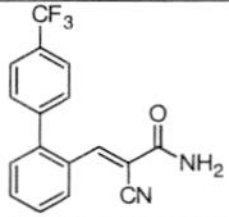 CF_3 O NH_2 CN	3.67	9.41	1	2	316.2824	302	14345	458	4196	168
40	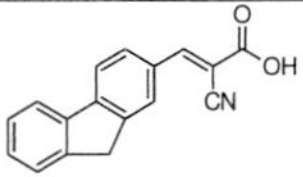 O OH CN	3.71	2.8	1	3	261.2799	342	34086	26951	33337	260.2

Nr	Structure	logP	pKa	HD	HA	Monoisotopic mass	λmax	ελmax	ελ355nm	ελ337 nm	MP °C
41	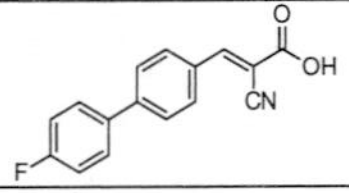	3.74	2.61	1	3	267.2593	323	28401	7340	23063	243.3
42	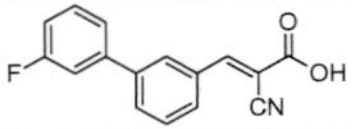	3.74	2.72	1	3	267.2593	285	16247	61	1064	188.1
43	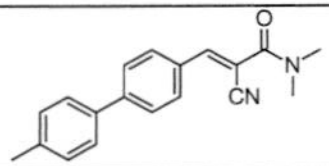	3.75	0	0	2	290.3648	333	28713	17949	28197	135.2
44	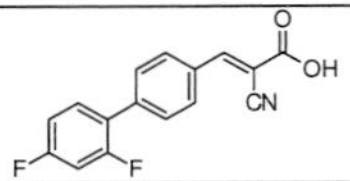	3.88	2.47	1	3	285.2498	315	22741	1895	12394	244.7
45	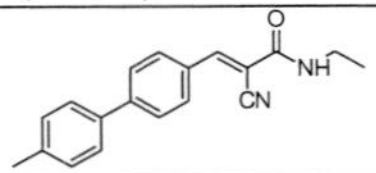	3.88	9.87	1	2	290.3648	340	28674	23879	28512	165.2
46	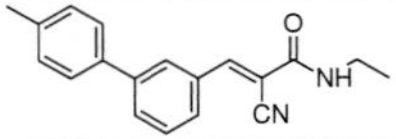	3.88	9.97	1	2	290.3648	301	20183	1311	3892	153.5
47	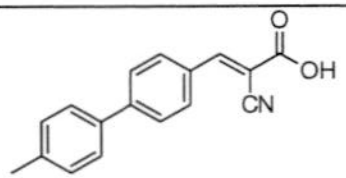	4.11	2.82	1	3	263.2958	331	28665	15262	27583	241.9
48	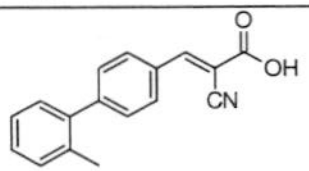	4.11	2.71	1	3	263.2958	309	19829	1385	7714	139.3

Nr	Structure	logP	pKa	HD	HA	Monoisotopic mass	λmax	ελmax	ελ355nm	ελ337 nm	MP °C
49	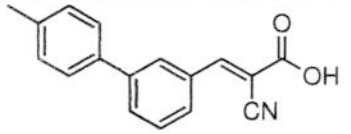	4.11	2.93	1	3	263.2958	263	24794	443	1822	176.7
50	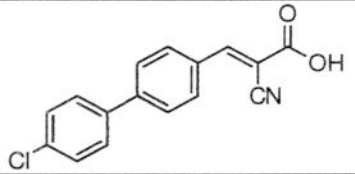	4.2	2.67	1	3	283.7139	324	32158	10062	27106	264.9
51	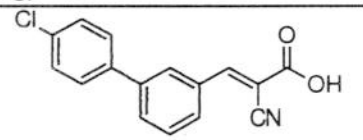	4.2	2.74	1	3	283.7139	262	28075	206	1432	204.4
52	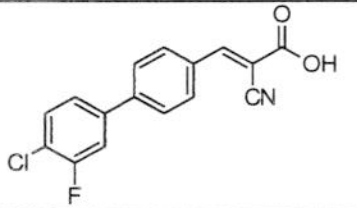	4.34	2.49	1	3	301.7044	329	28431	20725	28017	259.8
53	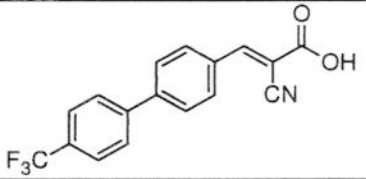	4.47	2.73	1	3	317.2671	318	31893	2651	19635	232.9
54	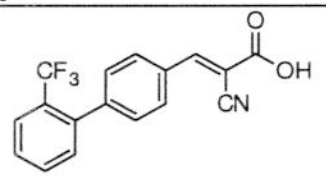	4.47	2.56	1	3	317.2671	302	22557	0	2446	152.3
55	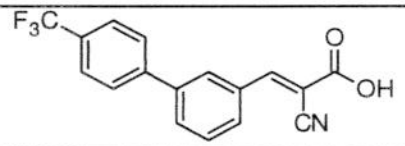	4.47	2.73	1	3	317.2671	257	25477	12	725	193.8
56	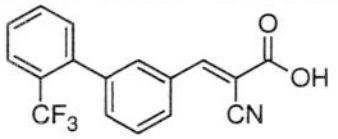	4.47	2.9	1	3	317.2671	270	2289	21	39	169.2

Nr	Structure	logP	pKa	HD	HA	Monoisotopic mass	λmax	ελmax	ελ355nm	ελ337 nm	MP °C
57	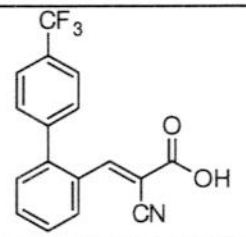 CF_3 O OH CN	4.47	2.71	1	3	317.2671	294	14501	470	1578	226.3
58	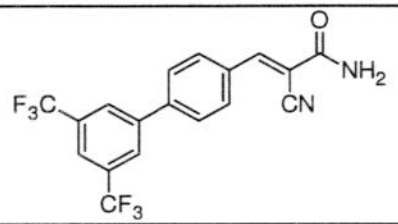 O NH_2 F_3C CN CF_3	4.54	9.98	1	2	384.2807	321	28978	4919	20898	217.2
59	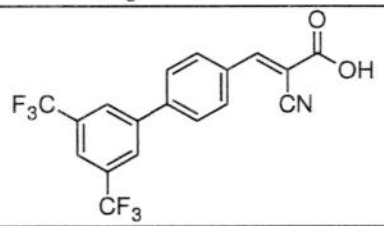 O OH F_3C CN CF_3	5.35	2.53	1	3	385.2654	312	27929	986	12086	227.5
60	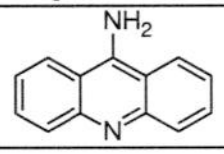 NH_2 N	2.68	9.29	1	2	194.2359	423	8437	1641	413	241*

Table1 Physical properties of used α-cyanocinnamic acid derivatives and 9-AA (Nr = matrix number, HD = number of hydogen bond donor atoms, HA = number of hydogen bond acceptor atoms, MP = melting point), *Melting point of matrix 60 from Sigma-Aldrich GmbH

References

[1] TANAKA, K., WAKI, H., IDO, Y., ET AL. **Protein and polymer analyses up to m/z 100 000 by laser ionization time-of-flight mass spectrometry**. *Rapid Communications in Mass Spectrometry*, **2**(8):151–153, 1988.

[2] KARAS, M. AND HILLENKAMP, F. **Laser desorption ionization of proteins with molecular masses exceeding 10,000 daltons**. *Analytical Chemistry*, **60**(20):2299–2301, 1988.

[3] BAHR, U., DEPPE, A., KARAS, M., HILLENKAMP, F., AND GIESSMANN, U. **Mass spectrometry of synthetic polymers by UV-matrix-assisted laser desorp-tion/ionization**. *Analytical Chemistry*, **64**(22):2866–2869, 1992.

[4] TABB, D. L., MCDONALD, W. H., AND YATES, J. R. **DTASelect and Contrast: tools for assembling and comparing protein identifications from shotgun proteomics**. *Journal of Proteome Research*, **1**(1):21–26, 2002.

[5] WU, K. J., STEDING, A., AND BECKER, C. H. **Matrix-assisted laser desorption time-of-flight mass spectrometry of oligonucleotides using 3-hydroxypicolinic acid as an ultraviolet-sensitive matrix**. *Rapid Communications in Mass Spectrometry*, **7**(2):142–146, 1993.

[6] HARVEY, D. J. **Matrix-assisted laser desorption/ionization mass spectrometry of carbohydrates**. *Mass Spectrometry Reviews*, **18**(6):349–450, 1999.

[7] KUSSMANN, M., NORDHOFF, E., RAHBEK-NIELSEN, H., ET AL. **Matrix-assisted laser desorption/ionization mass spectrometry sample preparation tech-niques designed for various peptide and protein analytes**. *Journal of Mass Spectrometry*, **32**(6):593–601, 1997.

[8] JUHASZ, P., COSTELLO, C. E., AND BIEMANN, K. **Matrix-assisted laser desorption ionization mass spectrometry with 2-(4-hydroxyphenylazo) benzoic acid matrix**. *Journal of the American Society for Mass Spectrometry*, **4**(5):399–409, 1993.

[9] CH'ANG, L.-Y., SCHELL, M., RINGELBERG, C., ET AL. **Detection of F508 mutation of the cystic fibrosis gene by matrix-assisted laser desorption/ionization mass spectrometry**. *Rapid Communications in Mass Spec-trometry*, **9**(9):772–774, 1995.

[10] TARANENKO, N. I., POT TER, N. T., ALLMAN, S. L., GOLOVLEV, V., AND CHEN, C. **Detection of trinucleotide expansion in neurodegenerative disease by matrix-assisted laser desorption/ionization time-of-flight mass spectrometry**. *Genetic Analysis: Biomolecular Engineering*, **15**(1):25–31, 1999.

[11] TRIMPIN, S., ROUHANIPOUR, A., AZ, R., RÄDER, H. J., AND MÜLLEN, K. **New aspects in matrix-assisted laser desorption/ionization time-of-flight mass spectrometry: a universal solvent-free sample preparation**. *Rapid Commu-nications in Mass Spectrometry*, **15**(15):1364–1373, 2001.

[12] PETRICOIN, E. F., ARDEKANI, A. M., HIT T, B. A., ET AL. **Use of proteomic patterns in serum to identify ovarian cancer**. *The Lancet*, **359**(9306):572– 577, 2002.

[13] FENNER, N. AND DALY, N. **Laser used for mass analysis**. *Review of Scientific Instruments*, **37**(8):1068–1070, 1966.

[14] VASTOLA, F. AND PIRONE, A. **Ionization of organic solids by laser irradiation**. *Advancements in Mass Spectrometry*, **4**:107–111, 1968.

[15] VASTOLA, F., MUMMA, R., AND PIRONE, A. **Analysis of organic salts by laser ionization**. *Organic Mass Spectrometry*, **3**(1):101–104, 1970.

[16] POSTHUMUS, M., KISTEMAKER, P., MEUZELAAR, H., AND TEN NOEVER DE BRAUW, M. **Laser desorption-mass spectrometry of polar nonvolatile bio-organic molecules**. *Analytical Chemistry*, **50**(7):985–991, 1978.

[17] JAGTAP, R. AND AMBRE, A. **Overview literature on matrix assisted laser desorption ionization mass spectroscopy (MALDI MS): basics and its applications in characterizing polymeric materials**. *Bulletin of Materials Sci-ence*, **28**(6):515–528, 2005.

[18] MACFARLANE, R. D. **Mass spectrometry of biomolecules: from PDMS to MALDI**. *Brazilian Journal of Physics*, **29**(3):415–421, 1999.

[19] TANAKA, K. **The origin of macromolecule ionization by laser irradiation (Nobel lecture)**. *Angewandte Chemie International Edition*, **42**(33):3860–3870, 2003.

[20] KARAS, M., BACHMANN, D., BAHR, U. E., AND HILLENKAMP, F. **Matrix-assisted ultraviolet laser desorption of non-volatile compounds**. *International Journal of Mass Spectrometry and Ion Processes*, **78**:53–68, 1987.

[21] KARAS, M., INGENDOH, A., BAHR, U., AND HILLENKAMP, F. **Ultraviolet–laser desorption/ionization mass spectrometry of femtomolar amounts of large proteins**. *Biomedical & Environmental Mass Spectrometry*, **18**(9):841– 843, 1989.

[22] BEAVIS, R. C. **Matrix-assisted ultraviolet laser desorption: Evolution and principles**. *Organic Mass Spectrometry*, **27**(6):653–659, 1992.

[23] NORDHOFF, E., INGENDOH, A., CRAMER, R., ET AL. **Matrix-assisted laser desorption/ionization mass spectrometry of nucleic acids with wavelengths in the ultraviolet and infrared**. *Rapid Communications in Mass Spectrometry*, **6**(12):771–776, 1992.

[24] OVERBERG, A., KARAS, M., BAHR, U., KAUFMANN, R., AND HILLENKAMP, F. **Matrix-assisted infrared-laser (2.94 μm) desorption/ionization mass spectrometry of large biomolecules**. *Rapid Communications in Mass Spectrometry*, **4**(8):293–296, 1990.

[25] OVERBERG, A., KARAS, M., HILLENKAMP, F., AND COT TER, R. **Matrix-assisted laser desorption of large biomolecules with a TEA-CO2-laser**. *Rapid Communications in Mass Spectrometry*, **5**(3):128–131, 1991.

[26] DREISEWERD, K., SCHÜRENBERG, M., KARAS, M., AND HILLENKAMP, F. **Influence of the laser intensity and spot size on the desorption of molecules and ions in matrix-assisted laser desorption/ionization with a uniform beam profile**. *International Journal of Mass Spectrometry and Ion Processes*, **141**(2):127–148, 1995.

[27] BERKENKAMP, S., MENZEL, C., HILLENKAMP, F., AND DREISEWERD, K. **Measurements of mean initial velocities of analyte and matrix ions in infrared matrix-assisted laser desorption ionization mass spectrometry**. *Journal of the American Society for Mass Spectrometry*, **13**(3):209–220, 2002.

[28] ZENOBI, R. AND KNOCHENMUSS, R. **Ion formation in MALDI mass spectrometry**. *Mass Spectrometry Reviews*, **17**(5):337–366, 1998.

[29] KARAS, M. AND KRÜGER, R. **Ion formation in MALDI: the cluster ionization mechanism**. *Chemical Reviews*, **103**(2):427–440, 2003.

[30] HILLENKAMP, F. AND KARAS, M. **Matrix-assisted laser desorption/ionisation, an experience**. *International Journal of Mass Spectrometry*, **200**(1):71–77, 2000.

[31] MENZEL, C., DREISEWERD, K., BERKENKAMP, S., AND HILLENKAMP, F. **Mechanisms of energy deposition in infrared matrix-assisted laser desorption/ionization mass spectrometry**. *International Journal of Mass Spectrometry*, **207**(1):73–96, 2001.

[32] DREISEWERD, K. **The desorption process in MALDI**. *Chemical Reviews*, **103**(2):395–426, 2003.

[33] WESTMACOTT, G., ENS, W., HILLENKAMP, F., DREISEWERD, K., AND SCHÜRENBERG, M. **The influence of laser fluence on ion yield in matrix-assisted laser desorption ionization mass spectrometry**. *International Journal of Mass Spectrometry*, **221**(1):67–81, 2002.

[34] MENZEL, C., DREISEWERD, K., BERKENKAMP, S., AND HILLENKAMP, F. **The role of the laser pulse duration in infrared matrix-assisted laser desorption/ionization mass spectrometry**. *Journal of the American Society for Mass Spectrometry*, **13**(8):975–984, 2002.

[35] JUHASZ, P., VESTAL, M., AND MARTIN, S. **On the initial velocity of ions generated by MALDI and its effects on the calibration of delayed extrac-tion TOF mass spectra**. *Journal of American Soceity for Mass Spectrometry*, **8**:209–217, 1997.

[36] KARAS, M., BAHR, U., FOURNIER, I., GLÜCKMANN, M., AND PFENNINGER, A. **The initial-ion velocity as a marker for different desorption-**

ionization mechanisms in MALDI. *International Journal of Mass Spectrometry*, **226**(1):239–248, 2003.

[37] HORNEFFER, V., DREISEWERD, K., LUDEMANN, H. C., ET AL. **Is the incorporation of analytes into matrix crystals a prerequisite for matrix-assisted laser desorption/ionization mass spectrometry? A study of five positional isomers of dihydroxybenzoic acid**. *International Journal of Mass Spectrometry*, **185**:859–870, 1999.

[38] GLÜCKMANN, M., PFENNINGER, A., KRÜGER, R., ET AL. **Mechanisms in MALDI analysis: surface interaction or incorporation of analytes?** *International Journal of Mass Spectrometry*, **210**:121–132, 2001.

[39] KNOCHENMUSS, R. **Ion formation mechanisms in UV-MALDI**. *Analyst*, **131**(9):966–986, 2006.

[40] EHRING, H., KARAS, M., AND HILLENKAMP, F. **Role of photoionization and photochemistry in ionization processes of organic molecules and relevance for matrix-assisted laser desorption lonization mass spectrometry**. *Organic Mass Spectrometry*, **27**(4):472–480, 1992.

[41] SPENGLER, B. AND BÖKELMANN, V. **Angular and time resolved intensity distributions of laser-desorbed matrix ions**. *Nuclear Instruments and Methods in Physics Research Section B: Beam Interactions with Materials and Atoms*, **82**(2):379–385, 1993.

[42] BÖKELMANN, V., SPENGLER, B., AND KAUFMANN, R. **Dynamical parameters of ion ejection and ion formation in matrix-assisted laser desorption/ionization**. *European Mass Spectrometry*, **27**:156–158, 1995.

[43] LIAO, P.C. AND ALLISON, J. **Ionization processes in matrix-assisted laser desorption/ionization mass spectrometry: Matrix-dependent formation of [M+ H]+ vs [M+ Na]+ ions of small peptides and some mechanistic comments**. *Journal of Mass Spectrometry*, **30**(3):408–423, 1995.

[44] GRIGOREAN, G., CAREY, R. I., AND AMSTER, I. J. **Studies of exchangeable protons in the matrix-assisted laser desorption/ionization process**. *European Mass Spectrometry*, **2**(2-3):139–143, 1996.

[45] LEHMANN, E., KNOCHENMUSS, R., AND ZENOBI, R. **Ionization mechanisms in matrix-assisted laser desorption/ionization mass spectrometry: contribution of preformed ions**. *Rapid Communications in Mass Spectrometry*, **11**(14):1483–1492, 1997.

[46] WONG, C. K., SO, M., AND CHAN, T. D. **Origins of the proton in the generation of protonated polymers and peptides in matrix-assisted laser desorption/ionization**. *European Mass Spectrometry*, **4**(3):223–232, 1998.

[47] KARBACH, V. AND KNOCHENMUSS, R. **Do single matrix molecules generate primary ions in ultraviolet matrix-assisted laser desorption/ionization**. *Rapid Communications in Mass Spectrometry*, **12**(14):968–974, 1998.

[48] MCCARLEY, T. D., MCCARLEY, R. L., AND LIMBACH, P. A. **Electron-transfer ionization in matrix-assisted laser desorption/ionization mass spectrometry**. *Analytical Chemistry*, **70**(20):4376–4379, 1998.

[49] LAND, C. AND KINSEL, G. **Investigation of the mechanism of intracluster proton transfer from sinapinic acid to biomolecular analytes**. *Journal of the American Society for Mass Spectrometry*, **9**(10):1060–1067, 1998.

[50] KNOCHENMUSS, R., KARBACH, V., WIESLI, U., BREUKER, K., AND ZENOBI, R. **The matrix suppression effect in matrix-assisted laser desorption/ionization: Application to negative ions and further characteristics**. *Rapid Communications in Mass Spectrometry*, **12**(9):529–534, 1998.

[51] KARAS, M., GLÜCKMANN, M., AND SCHÄFER, J. **Ionization in matrix-assisted laser desorption/ionization: singly charged molecular ions are the lucky survivors**. *Journal of Mass Spectrometry*, **35**(1):1–12, 2000.

[52] FRANKEVICH, V., KNOCHENMUSS, R., AND ZENOBI, R. **The origin of electrons in MALDI and their use for sympathetic cooling of negative ions in FTICR**. *International Journal of Mass Spectrometry*, **220**(1):11–19, 2002.

[53] BREUKER, K., KNOCHENMUSS, R., ZHANG, J., STORTELDER, A., AND ZENOBI, R. **Thermodynamic control of final ion distributions in MALDI: in-plume proton transfer reactions**. *International Journal of Mass Spectrometry*, **226**(1):211–222, 2003.

[54] FOURNIER, I., MARINACH, C., TABET, J., AND BOLBACH, G. **Irradiation effects in MALDI, ablation, ion production, and surface modifications. Part II: 2, 5-dihydroxybenzoic acid monocrystals**. *Journal of the American Society for Mass Spectrometry*, **14**(8):893–899, 2003.

[55] SYAGE, J. A. **Mechanism of [M+ H]+ formation in photoionization mass spectrometry**. *Journal of the American Society for Mass Spectrometry*, **15**(11):1521–1533, 2004.

[56] KNOCHENMUSS, R. **A quantitative model of ultraviolet matrix-assisted laser desorption/ionization**. *Journal of Mass Spectrometry*, **37**(8):867–877, 2002.

[57] CHEN, X., CARROLL, J. A., AND BEAVIS, R. C. **Near-ultraviolet-induced matrix-assisted laser desorption/ionization as a function of wavelength**. *Journal of the American Society for Mass Spectrometry*, **9**(9):885–891, 1998.

[58] NIU, S., ZHANG, W., AND CHAIT, B. T. **Direct comparison of infrared and ultraviolet wavelength matrix-assisted laser desorption/ionization mass spectrometry of proteins**. *Journal of the American Society for Mass Spectrometry*, **9**(1):1–7, 1998.

[59] MOON, J. H., SHIN, Y. S., BAE, Y. J., AND KIM, M. S. **Ion yields for some salts in MALDI: mechanism for the gas-phase ion formation from preformed ions**. *Journal of the American Society for Mass Spectrometry*, **23**(1):162–170, 2012.

[60] MOON, J. H., YOON, S. H., AND KIM, M. S. **Temperature of peptide ions generated by matrix-assisted laser desorption ionization and their dissociation kinetic parameters**. *The Journal of Physical Chemistry B*, **113**(7):2071–2076, 2009.

[61] BAE, Y. J., SHIN, Y. S., MOON, J. H., AND KIM, M. S. **Degree of ionization in MALDI of peptides: thermal explanation for the gas-phase ion formation**. *Journal of the American Society for Mass Spectrometry*, **23**(8):1326–1335, 2012.

[62] LAI, Y.H., WANG, C.C., LIN, S.H., LEE, Y. T., AND WANG, Y.S. **Solid-phase thermodynamic interpretation of ion desorption in matrix-**

assisted laser desorption/ionization. *The Journal of Physical Chemistry B*, **114**(43):13847–13852, 2010.

[63] TRIMPIN, S., WANG, B., INUTAN, E. D., ET AL. **A mechanism for ionization of nonvolatile compounds in mass spectrometry: considerations from MALDI and inlet ionization**. *Journal of the American Society for Mass Spectrometry*, **23**(10):1644–1660, 2012.

[64] VESTAL, M. L. **Ionization techniques for nonvolatile molecules**. *Mass Spectrometry Reviews*, **2**(4):447–480, 1983.

[65] HOTELING, A. J., ERB, W. J., TYSON, R. J., AND OWENS, K. G. **Exploring the Importance of the Relative Solubility of Matrix and Analyte in MALDI Sample Preparation Using HPLC Exploring the Importance of the Relative Solubility of Matrix and Analyte in MALDI Sample Preparation Using HPLC**. *Analytical Chemistry*, **76**(17):5157–5164, 2004.

[66] LEWIS, J. K., WEI, J., AND SIUZDAK, G. **Matrix-assisted Laser Desorption / Ionization Mass Spectrometry in Peptide and Protein Analysis**. *Encyclopedia of Analytical Chemistry*, pages 5880–5894, 2000.

[67] VORM, O., ROEPSTORFF, P., AND MANN, M. **Improved resolution and very high sensitivity in MALDI TOF of matrix surfaces made by fast evaporation**. *Analytical Chemistry*, **66**(19):3281–3287, 1994.

[68] VORM, O. AND MANN, M. **Improved mass accuracy in matrix-assisted laser desorption/ionization time-of-flight mass spectrometry of peptides**. *Journal of the American Society for Mass Spectrometry*, **5**(11):955–958, 1994.

[69] COHEN, S. L. AND CHAIT, B. T. **Influence of matrix solution conditions on the MALDI-MS analysis of peptides and proteins**. *Analytical Chemistry*, **68**(1):31–37, 1996.

[70] XIANG, F. AND BEAVIS, R. C. **Growing protein-doped sinapic acid crystals for laser desorption: An alternative preparation method for difficult samples**. *Organic Mass Spectrometry*, **28**(12):1424–1429, 1993.

[71] GOODWIN, R. J. A., PENNINGTON, S. R., AND PIT T, A. R. **Protein and peptides in pictures: Imaging with MALDI mass spectrometry**. *Proteomics*, **8**(18):3785–3800, 2008.

[72] GUSTAFSSON, J. O. R., OEHLER, M. K., RUSZKIEWICZ, A., MCCOLL, S. R., AND HOFFMANN, P. **MALDI imaging mass spectrometry (MALDI-IMS)-application of spatial proteomics for ovarian cancer classification and diagnosis**. *International Journal of Molecular Sciences*, **12**(1):773–794, 2011.

[73] WENK, M. R. **The emerging field of lipidomics**. *Nature Reviews Drug Discovery*, **4**(7):594–610, 2005.

[74] WOODS, A. S., WANG, H.Y. J., AND JACKSON, S. N. **A snapshot of tissue glycerolipids**. *Current Pharmaceutical Design*, **13**(32):3344–3356, 2007.

[75] PIOMELLI, D. **The challenge of brain lipidomics**. *Prostaglandins & other Lipid mediators*, **77**(1):23–34, 2005.

[76] WOODS, A. S. AND JACKSON, S. N. **Brain tissue lipidomics: Direct probing using matrix-assisted laser desorption/ionization mass spectrometry**. *The AAPS Journal*, **8**(2):E391–E395, 2006.

[77] FUJIWAKI, T., YAMAGUCHI, S., SUKEGAWA, K., AND TAKETOMI, T. **Application of delayed extraction matrix-assisted laser desorption ionization time-of-flight mass spectrometry for analysis of sphingolipids in tissues from sphingolipidosis patients**. *Journal of Chromatography B: Biomedical Sciences and Applications*, **731**(1):45–52, 1999.

[78] HE, X., CHEN, F., MCGOVERN, M. M., AND SCHUCHMAN, E. H. **A fluorescence-based, high-throughput sphingomyelin assay for the analysis of Niemann–Pick disease and other disorders of sphingomyelin metabolism**. *Analytical Biochemistry*, **306**(1):115–123, 2002.

[79] HAN, X., M HOLTZMAN, D., W MCKEEL, D., KELLEY, J., AND MORRIS, J. C. **Substantial sulfatide deficiency and ceramide elevation in very early Alzheimer's disease: potential role in disease pathogenesis**. *Journal of Neurochemistry*, **82**(4):809–818, 2002.

[80] MURPHY, E. J., SCHAPIRO, M. B., RAPOPORT, S. I., AND SHET TY, H. U. **Phospholipid composition and levels are altered in Down syndrome brain**. *Brain Research*, **867**(1):9–18, 2000.

[81] WOODS, A. S., MOYER, S. C., WANG, H.Y. J., AND WISE, R. A. **Interaction of chlorisondamine with the neuronal nicotinic acetylcholine receptor**. *Journal of Proteome Research*, **2**(2):207–212, 2003.

[82] WOODS, A. S., UGAROV, M., EGAN, T., ET AL. **Lipid/peptide/nucleotide separation with MALDI-ion mobility-TOF MS**. *Analytical Chemistry*, **76**(8):2187–2195, 2004.

[83] SUZUKI, K. ET AL. **Chemistry and metabolism of brain lipids**. *Basic Neurochemistry*, **1**:207–227, 1981.

[84] SONNINO, S. AND CHIGORNO, V. **Ganglioside molecular species containing C18-and C20-sphingosine in mammalian nervous tissues and neuronal cell cultures**. *Biochimica et Biophysica Acta (BBA)-Reviews on Biomembranes*, **1469**(2):63–77, 2000.

[85] PETKOVIC, M., SCHILLER, J., MÜLLER, M., ET AL. **Detection of individual phospholipids in lipid mixtures by matrix-assisted laser desorption/ionization time-of-flight mass spectrometry: phosphatidylcholine prevents the detection of further species**. *Analytical Biochemistry*, **289**(2):202–216, 2001.

[86] SCHILLER, J., ARNHOLD, J., BENARD, S., ET AL. **Lipid analysis by matrix-assisted laser desorption and ionization mass spectrometry: a methodological approach**. *Analytical Biochemistry*, **267**(1):46–56, 1999.

[87] SCHILLER, J., SÜSS, R., ARNHOLD, J., ET AL. **Matrix-assisted laser desorption and ionization time-of-flight (MALDI-TOF) mass spectrometry in lipid and phospholipid research**. *Progress in Lipid Research*, **43**(5):449–488, 2004.

[88] NIMESH, S., MOHOT TALAGE, S., VINCENT, R., AND KUMARATHASAN, P. **Current status and future perspectives of mass spectrometry imaging**. *International Journal of Molecular Sciences*, **14**(6):11277–11301, 2013.

[89] CERRUTI, C. D., BENABDELLAH, F., LAPRÈVOTE, O., TOUB OUL, D., AND BRUNELLE, A. **MALDI imaging and structural analysis of rat brain lipid negative ions with 9-aminoacridine matrix**. *Analytical Chemistry*, **84**(5):2164–2171, 2012.

[90] WOODS, A. S. AND JACKSON, S. N. **Brain tissue lipidomics: direct probing using matrix-assisted laser desorption/ionization mass spectrometry.** *The AAPS journal*, **8**(2):E391–E395, 2006.

[91] WANG, H. Y. J., POST, S. N. J. J., AND WOODS, A. S. **A minimalist approach to MALDI imaging of glycerophospholipids and sphingolipids in rat brain sections**. *International Journal of Mass Spectrometry*, **278**(2-3):143– 149, 2008.

[92] JACKSON, S. N., WANG, H. Y. J., AND WOODS, A. S. **In situ structural characterization of phosphatidylcholines in brain tissue using MALDI-MS/MS**. *Journal of the American Soceity for Mass Spectrometry*, **16**(12):2052-2056, 2005

[93] JACKSON, S. N., WANG, H. Y. J., AND WOODS, A. S. **Direct profiling of lipid distribution in brain tissue using MALDI-TOFMS**. *Analytical Chem-istry*, **77**(14):4523–4527, 2005.

[94] KRAUSE, J., STOECKLI, M., AND SCHLUNEGGER, U. P. **Studies on the selection of new matrices for ultraviolet matrix-assisted laser desorption/ionization time-of-flight mass spectrometry**. *Rapid Communications in Mass Spectrometry*, **10**(15):1927–1933, 1996.

[95] CERRUTI, C. D., BENABDELLAH, F., LAPREÌ A̧VOTE, O., TOUB OUL, D., AND BRUNELLE, A. **MALDI imaging and structural analysis of rat brain lipid negative ions with 9-aminoacridine matrix**. *Analytical Chemistry*, **84**(5):2164–2171, 2012.

[96] TEUBER, K., SCHILLER, J., FUCHS, B., KARAS, M., AND JASKOLLA, T. W. **Significant sensitivity improvements by matrix optimization: A MALDI-TOF mass spectrometric study of lipids from hen egg yolk**. *Chemistry and Physics of Lipids*, **163**(6):552–560, 2010.

[97] FÜLÖP, A., PORADA, M. B., MARSCHING, C., ET AL. **4-phenyl-α-cyanocinnamic acid amide: Screening for a negative ion matrix for MALDI-MS imaging of multiple lipid classes**. *Analytical Chemistry*, **85**(19):9156–9163, 2013.

[98] NONAMI, H., WU, F., THUMMEL, R. P., ET AL. **Evaluation of pyridoin-doles, pyridylindoles and pyridylpyridoindoles as matrices for ultraviolet matrix-assisted laser desorption/ionization time-of-flight mass spectrom-etry**. *Rapid Communications in Mass Spectrometry*, **15**(23):2354–2373, 2001.

[99] PORTA, T., GRIVET, C., KNOCHENMUSS, R., VARESIO, E., AND HOPFGARTNER, G. **Alternative CHCA-based matrices for the analysis of low molecular weight compounds by UV-MALDI-tandem mass spectrometry**. *Journal of Mass Spectrometry*, **46**(2):144–152, 2011.

[100] JASKOLLA, T. W., LEHMANN, W.D., AND KARAS, M. **4-Chloro-α-cyanocinnamic acid is an advanced, rationally designed MALDI matrix.** *Proceedings of the National Academy of Sciences of the United States of America*, **105**(34):12200–12205, 2008.

[101] DIPPY, J. F. J. AND EVANS, R. M. **THE NATURE OF THE CATALYST IN THE PERKIN CONDENSATION**. *The Journal of Organic Chemistry*, **15**(3):451–456, 1950.

[102] SHARMA, P. **Cinnamic acid derivatives: A new chapter of various pharmacological activities**. *Journal of Chemical and Pharmaceutical Research*, **3**(2):403–23, 2011.

[103] BUCKLES, R. E. AND HAUSMAN, E. A. **A Simplified Synthesis of α-Phenylcinnamic Acid and α-Phenyl-p-nitrocinnamic Acid1**. *Journal of the American Chemical Society*, **70**(1):415–415, 1948.

[104] KNOEVENAGEL, E. **Condensation von Malonsäure mit aromatischen Aldehyden durch Ammoniak und Amine**. *Berichte der Deutschen Chemischen Gesellschaft*, **31**(3):2596–2619, 1898.

[105] WILEY, R. H. AND SMITH, N. **Nitrostyrenes and 2-Nitro-5-vinylfuran1**. *Journal of the American Chemical Society*, **72**(11):5198–5199, 1950.

[106] HECK, R. F. AND NOLLEY JR, J. **Palladium-catalyzed vinylic hydrogen substitution reactions with aryl, benzyl, and styryl halides**. *The Journal of Organic Chemistry*, **37**(14):2320–2322, 1972.

[107] MIZOROKI, T., MORI, K., AND OZAKI, A. **Arylation of olefin with aryl iodide catalyzed by palladium**. *Bulletin of the Chemical Society of Japan*, **44**(2):581– 581, 1971.

[108] AMBULGEKAR, G. V., BHANAGE, B. M., AND SAMANT, S. D. **Low temperature recyclable catalyst for Heck reactions using ultrasound**. *Tetrahedron Letters*, **46**(14):2483–2485, 2005.

[109] ZHANG, Z., ZHA, Z., GAN, C., ET AL. **Catalysis and regioselectivity of the aqueous Heck reaction by Pd (0) nanoparticles under ultrasonic irradiation**. *The Journal of Organic Chemistry*, **71**(11):4339–4342, 2006.

[110] RAO, Y. S. AND FILLER, R. **trans-to cis-Isomerisation of cinnamic acids and related carbonyl compounds**. *Journal of the Chemical Society, Chemical Communications*, (12):471–472, 1976.

[111] HOCKING, M. B. **Photochemical and thermal isomerizations of cis-and trans-cinnamic acids, and their photostationary state**. *Canadian Journal of Chemistry*, **47**(24):4567–4576, 1969.

[112] LINDLAR, H. **Ein neuer Katalysator für selektive Hydrierungen**. *Helvetica Chimica Acta*, **35**(2):446–450, 1952.

[113] WADSWORTH, W. S. **Synthetic Applications of Phosphoryl-Stabilized Anions**. *Organic Reactions*, 1977.

[114] HORNER, L., HOFFMANN, H., AND WIPPEL, H. G. **Phosphororganische verbindungen, XII. Phosphinoxyde als olefinierungsreagenzien**. *Chemische Berichte*, **91**(1):61–63, 1958.

[115] WADSWORTH, W. S. AND EMMONS, W. D. **The utility of phosphonate carbanions in olefin synthesis**. *Journal of the American Chemical Society*, **83**(7):1733–1738, 1961.

[116] MICHAELIS, A. AND KAEHNE, R. **Ueber das Verhalten der Jodalkyle gegen die sogen. Phosphorigsäureester oder O-Phosphine**. *Berichte der deutschen chemischen Gesellschaft*, **31**(1):1048–1055, 1898.

[117] BHAT TACHARYA, A. K. AND THYAGARAJAN, G. **Michaelis-arbuzov rearrangement**. *Chemical Reviews*, **81**(4):415–430, 1981.

[118] HASWELL, S. J. **Introduction to chemometrics**. *Practical Guide to Chemometrics. New York: Marcel Dekker Inc, USA*, pages 1–3, 1992.

[119] HILARIO, M. AND KALOUSIS, A. **Approaches to dimensionality reduction in proteomic biomarker studies**. *Briefings in Bioinformatics*, **9**(2):102–118, 2008.

[120] HILARIO, M., KALOUSIS, A., PELLEGRINI, C., AND MUELLER, M. **Processing and classification of protein mass spectra**. *Mass Spectrometry Reviews*, **25**(3):409–449, 2006.

[121] LEE, K. R., LIN, X., PARK, D. C., AND ESLAVA, S. **Megavariate data analysis of mass spectrometric proteomics data using latent variable projection method**. *Proteomics*, **3**(9):1680–1686, 2003.

[122] WISE, B. M., GALLAGHER, N. B., BRO, R., ET AL. **PLS_Toolbox 3.5 for use with MATLABâD ´c. Eigenvector Research**. *Inc, Wenatchee, WA*, 2005.

[123] JASKOLLA, T. W. AND KARAS, M. **Using Fluorescence Dyes as a Tool for Analyzing the MALDI Process**. *Journal of the American Society for Mass Spectrometry*, **19**(8):1054–1061, 2008.

[124] KNOCHENMUSS, R., DUB OIS, F., DALE, M. J., AND ZENOBI, R. **The matrix suppression effect and ionization mechanism in MALDI**. *Rapid Commun Mass Spectrom*, **10**:871, 1996.

[125] FÜLLÖP, A., PORADA, M. B., MARSCHING, C., ET AL. **4-Phenyl-α-cyanocinnamic acid amide: screening for a negative ion matrix for MALDI-MS imaging of multiple lipid classes**. *Analytical Chemistry*, **85**(19):9156–9163, 2013.

[126] MARSCHING, C., ECKHARDT, M., GRÖNE, H.-J., SANDHOFF, R., AND HOPF, C. **Imaging of complex sulfatides SM3 and SB1a in mouse kidney using MALDI-TOF/TOF mass spectrometry**. *Analytical and Bioanalytical Chemistry*, **401**(1):53–64, 2011.

[127] JACKSON, S. N., WANG, H.-Y. J., AND WOODS, A. S. **In Situ Structural Characterization of Phosphatidylcholines in Brain Tissue Using MALDI-MS/MS**. *Journal of the American Society for Mass Spectrometry*, **16**(12):2052 – 2056, 2005.

[128] JACKSON, S. N., WANG, H. Y. J., AND WOODS, A. S. **In Situ Structural Characterization of Glycerophospholipids and Sulfatides in Brain Tissue Using MALDI-MS/MS**. *Journal of the American Society for Mass Spectrometry*, **18**(1):17–26, 2007.

[129] Soltwisch, J., Jaskolla, T. W., Hillenkamp, F., Karas, M., and Dreisewerd, K. **Ion yields in UV-MALDI mass spectrometry as a function of excitation laser wavelength and optical and physico-chemical properties of classical and halogen-substituted MALDI matrixes**. *Analytical Chemistry*, **84**(15):6567–6576, 2012.

[130] Allwood, D., Dreyfus, R., Perera, I., and Dyer, P. **UV Optical Absorption of Matrices Used for Matrix-assisted Laser Desorption/Ionization**. *Rapid Communications in Mass Spectrometry*, **10**(13):1575–1578, 1996.

Acknowledgements

I would like to express my deepest appreciation to main supervisor Prof. Dr. Stefan Bräse for giving me an opportunity to conduct the doctoral research under his expert guidance. It is truly an honor to be a part of his research team at the Karlsruhe Institute of Technology, Germany. I also convey my sincere gratitude to Prof. Dr. Hans-Dieter Junker (University of Applied Sciences, Aalen) for the consistent support during my PhD, for his patience, motivation, and impeccable knowledge. His guidance helped me in all the time of research and writing of this thesis.

I sincerely thank Prof. Dr. Burkhard Luy, Prof. Dr. Annie Powell and Prof. Dr. Christopher Barner-Kowollik of KIT, Germany for reviewing my thesis work.

I also thank Prof. Carsten Hopf (University of Applied Sciences, Mannheim) for allowing me to be a part of the larger project on MALDI-MS and providing his expert guidance whenever required. Furthermore, an important aspect of my thesis was carrying out measurements in Mannheim and I thank Dr. Annabelle Fülöp very much for her great support regarding the same.

I highly appreciate Prof. Dr. Dirk Flottmann's (University of Applied Sciences, Aalen) support especially for guidance related to statistical analysis. My earnest thanks to Mr. Andreas Haible, Mr. Peter Pfundstein, who gave me the access to laboratory facilities and for promptly providing me the necessary chemicals from time to time and Mr. Silvin Hippich for technical support and for providing all the required software. Without their judicious support it would have been impossible to conduct this research.

I am equally grateful to my fellow lab-mates Henning Blott, Nora Tromsdorf and Martina Porada for stimulating discussions, providing feedbacks, spending long hours with me for measurements and all the fun we have had in last 5 years.

Last but not the least; I feel obliged to my parents Mr. Bhaskar Tambe and Mrs. Shubhalaxmi Tambe, my mother-in-law Ms. Shubhangi Deshpande for everything they have done for me so far. They have been an enormous support and I am thoroughly indebted. I finally thank my wife Dr. Shalaka Shah for her precious support during PhD, for motivating me time and again, for making this PhD journey enjoyable and generally for the last 9 years of togetherness.

Curriculum Vitae

Suparna Bhaskar Tambe
Date of Birth: 26.09.1977 in Satara, India
Nationality: Indian
E-mail: suparna.tambe@gmail.com

EDUCATIONAL QUALIFICATIONS

10/2011-12/2016	**Doctoral Thesis** under the guidance of Prof. Dr. Stefan Bräse, Institute of Organic Chemistry, Karlsruhe Institute of Technology (KIT), Germany Topic: *Structure Performance Relationships of the Novel MALDI-MS Matrices*
10/2008-04/2011	**Master of Science**, Technical University of Braunschweig, Germany
09/2010-02/2011	**Master Thesis** Under the guidance of Dr. Jeroen Dickschat and Prof. Dr. Stefan Schulz, Institute for Organic Chemistry, Technical University of Braunschweig, Germany Topic: *Investigations in Synthesis and Bio-synthesis of Tropodithietic* Acid
06/2001-05/2003	**Master of Science**, University of Pune, India
09/2002-02/2003	**Master Thesis** under the guidance of Prof. Dr. Rajashree Kashalkar, Sir Parshurambhau College, University of Pune, India Topic: *Characterisation of Water and Alcohol Soluble Components in Medicinal Plants*
06/1995-04/2000	**Bachelor of Science**, University of Pune, India
06/1994-04/1995	Higher Secondary School Certificate Examination (**12th grade**), Kolhapur Board, India
06/1992-04/1993	Secondary School Certificate Examination (**10th grade**), Kolhapur Board, India

WORK EXPERIENCE

06/2006–08/2008	Visiting Lecturer, Institute of Environment Education and Research, Bharati Vidyapeeth University, Pune, India
06/2003–05/2006	Visiting Lecturer, Sir Parashurambhau College, University of Pune, India
06/2003-05/2005	Visiting Lecturer, Vishwakarma High School and Junior College, Pune, India
06/2003-05/2004	Laboratory Demonstrator, Fergusson College, University of Pune, India

List of Publications

1) Fülöp A, Porada MB, Marsching C, Blott H, Meyer B, Tambe S, Sandhoff R, Junker HD, Hopf C; **4-Phenyl-alpha-cyanocinnamic acid amide: screening for a negative ion matrix for MALDI-MS imaging of multiple lipid classes**. *Analytical Chemistry*, 2013, **85**(19):9156–9163.

2) Tambe S, Blott H, Fülöp A, Spang N , Flottmann D, Bräse S, Hopf C, Junker HD; **Structure Performance Relationships of Phenyl Cinncamic Acid Derivatives as MALDI-MS Matrices for Sulfatide Detection**. *Analytical and Bioanalytical Chemistry*, 2016, **409**(6):1569-1580.